Stanley Baron

THE
DESERT
LOCUST

EYRE METHUEN · LONDON

First published in Great Britain 1972
© 1972 Stanley Baron
Printed in Great Britain for
Eyre Methuen Ltd
11 New Fetter Lane, EC4P 4EE
by T. & A. Constable Ltd
Hopetoun Street, Edinburgh EH7 4NF

SBN 413 27150 1

The Desert Locust

Contents

Illustrations

MAPS

Acknowledgement and thanks for permission to reproduce the plates is due to the FAO, G. Tortoli, for plates 2, 3, 4a, 6 and 15b; to the ALRC for plates 4b, 12b and 15a; to the FAO, P. Keen, for plates 7a and 7b; to Clifford Ashall for plate 8; to the FAO, Jean Manuel, for plate 11 and to Shell for the back cover picture.

The maps were prepared by Neil Hyslop. Data was supplied by the Anti-Locust Research Centre.

Preface and Acknowledgements

The Locust of my title is *Schistocerca gregaria*, the Desert Locust, a creature unique among the fourteen or so species of its kind in having no permanent outbreak areas. This fact, in itself, long bedevilled hopes of control; but we now know that among the limitless sand seas of the Near and Middle East it has certain preferred types of habitat where vegetation has been able to spring to life following rain. Since this may occur but rarely, *Schistocerca* has developed highly nomadic habits. No other insect in the world travels so far, over such inhospitable country, merely in order to keep alive, and none is so swiftly able to expand both its numbers and its territory when conditions develop in its favour.

For many years the Anti-Locust Research Centre, a British government organization, fostered campaigns against the Desert Locust in Commonwealth countries periodically plagued by it. The Centre still plays a vital role, but now works in conjunction with the Food and Agriculture Organization of the United Nations, by which all the campaigns, both regional and national, in more than forty countries extending from the West Atlantic to the Bay of Bengal, are coordinated. The Anti-Locust Research Centre, to which I shall continue to refer, usually by its initials, has recently been reorganized and amalgamated with other British pest research units, under a new title – Centre for Overseas Pest Research.

Not all their activities are described here, for chasing a wild

ix

goose is as nothing compared with the time and effort exacted in the search for the Desert Locust in its non-gregarious phase, when one may travel thousands of miles and find almost nothing. These quiescent periods are, nevertheless, of as great importance to the scientist as the plague periods. It was during one of them, sitting literally at the feet – beside a Sahara camp fire – of Professor Roger Pasquier of the School of Agriculture of Algiers, that I received my first lessons in the environmental factors in the Desert Locust's life and behaviour.

Pasquier is one of a comparatively small group of locust experts whose field researches have been continuous over many years. Others to whom I am especially indebted are Dr R. C. Rainey, who contributed much to my understanding of the meteorological factors, and his colleagues in the ALRC, Jeremy Roffey and George Popov, whose eye-witness accounts of all the stages of a recent West African outbreak I have quoted in length. All three, together with Christopher Hemming of the ALRC, and Professor Pasquier, have read appropriate parts of my manuscript and where necessary have suggested changes.

In Ethiopia my mentor was John Sayer, chief scientist of the Desert Locust Control Organization of Eastern Africa. I am grateful to him and to the DLCO's Director, Adefris Bellehu, for some of my most exhilarating, if unnerving, flying experiences during their control operations.

I have had, of course, to consult innumerable publications, and two of the most thumbed are Miss Zena Waloff's invaluable biogeographical analysis – *Anti-Locust Memoir No. 8* – of the major upsurges, plagues and recessions of this century, and the ALRC's *Locust Handbook*, a model of compressed information whose chief editor, Mr Clifford Ashall, is in charge of the Centre's field activities. He has always been ready to straighten me out when I have been in doubt or difficulty and I am grateful to him for having read my proofs. Other published sources are as a rule mentioned in the text. The manifold scientific papers dealing with various aspects of this most-discussed of insects are available to students by arrangement with the Librarian, Centre for Overseas Pest Research. I have not, therefore, included a bibliography.

The most important source of information, Sir Boris Uvarov,

founder and first Director of the ALRC, is, alas, no longer alive. His great work lives on in the first volume of *Grasshoppers and Locusts*, published just before his death; the second volume, on which he was then working, will be finished by other hands. There can be few entomologists to whom such a large part of the world's population owes so much, equally for his genius as a practical theoretician – it was he who first realized that the apparently harmless locust of the recession and the frightful creature of the plague were the same species – and for his organizational powers in the campaigns of well over forty years.

His successor as Director, Dr Peter Haskell, has since also helped to guide my rather groping steps.

Finally I must give thanks to my friend and sometime colleague Mr Gurdas Singh, Chief Locust Officer of FAO, who set me off on my quest and has subsequently had to pay for it in my endless questionings.

Introduction

In January 1966 I found myself travelling in a Landrover across the Sahara, from Tamanrasset, southernmost of the Algerian oases, to a remote wadi called Wadi In Attenkarer close to the point where the borders of Algeria, Mali and Niger meet; then north again and east past the spurs of the Hoggar Mountains, watching them change from black and blue to orange, to the modestly developing tourist centre of Djanet (famous for its rock paintings), where my journey ended on the border of Libya – a border just as invisible as the others had been.

The object of this excursion, which took about ten days and was followed by further, similar journeys through some of the wilds of Saudi Arabia and Iran, was to report on the progress of the multi-million-dollar Desert Locust Project launched six years earlier by the Food and Agriculture Organization of the United Nations (FAO) with the financial backing of the UN Development Programme (or Special Fund, as it was then called).

I had been working for FAO for some months past, but knew little at this time of the habits of the most historic of the world's insect pests. Nor did it seem likely that I would learn much more in 1966, when the locust – a creature given to extremes – was in such complete recession that it would have been hard to find more than a few thousand in the five million square miles or so of desert containing its natural habitats. Thanks to the unceasing encouragement and help of scientists who became my friends, I found myself – a layman – engrossed by the subject of this extraordinary insect whose name is synonymous with famine among one-eighth of the world's people. Having witnessed some of the hostilities vented on it by nature between plagues, I also developed a sneaking sympathy with it, whether found in the wilds or in that unlikely habitat, the laboratories of the Anti-Locust Research Centre, located behind the fashion stores of London's Kensington High Street, where I also spent much time.

Then, in the winter of 1967-8, new outbreaks in Africa and Arabia brought a fresh threat of plague and I was able to go to Ethiopia and Sudan to watch the battle for control. Thanks to the efforts of dedicated experts, often living and working under conditions of great discomfort and considerable danger, this invasion was stopped. Whether this means final defeat for the locust may depend, in the long run, on the ability and willingness of some forty governments to continue the cooperation it has forced upon them.

Chapter 1

The Locust in History

Since the beginnings of recorded history locusts have been a terror
to men of a kind and on a scale not known in the case of any other
pest. The ability of the insect to emerge seemingly from nowhere
and to multiply so enormously that the soil where it bred appeared
to erupt hopping nymphs as far as the eye could see; the joining
together of the hoppers in vast marching bands apparently in-
destructible by any other creature – all these aspects of its life
cycle combined, even before the flying swarms appeared, to
produce awe and superstitious horror in its victims. When the
plague fell on cultivators and their crops without warning, as it
often did, the catastrophe assumed a sense of doom far greater
than that provoked by any human invasion. As the peasant worked
in his field he saw the swarm approaching from an empty landscape. At first it seemed like smoke on the horizon billowing from
an unseen bush fire, sometimes puffing up into pillars and spirals,
then subsiding and thickening just above the skyline into an advancing aerial horde so wide and dense that the sun vanished.

We know from modern measurements of swarm areas and
volumes that the descriptions, repeatedly given in the Bible and
elsewhere, of the sky being darkened and the sun eclipsed, are
literally correct. For instance, during the plague that continued
from 1948 to 1963, several swarms were recorded as exceeding
a hundred square miles; and one is said to have been the size
of London. The mighty descriptive book of *Joel* (page 119) is

1

inaccurate only in one respect. Having listed the depredations of a visitation of locusts, so that we can almost see the laying waste of the vines and the splintering of the fig trees, and hear the firelike crackle of the advancing insects before they leap upon the city, run upon the walls, climb up into the houses and enter through the windows, Joel concludes: 'Their like has never been from of old, nor will be again after them through the years of all generations.' Today we can say with reasonable certainty that during no period from Joel's day to this have any of the recessions between major plagues lasted more than a few years. Five such plagues have occurred this century and the longest lull between them was seven years. Out of sixty-one years since 1910, some forty have been plague years. From 1949 to 1963 the plague raged back and forth virtually unchecked; then came a five-year recession followed by yet another outbreak calling for the international control campaign I describe in this book.

There is no reason to believe that matters were any different in Biblical times. Indeed, the very consistency with which plagues of locusts were held up to man as examples of the very worst fate that could befall him testify to their frequency. The Eighth Plague of Egypt, described in *Exodus*, is presumed to have visited that country 3,500 years ago. Images of locusts were carved on Sixth-Dynasty tombs at Saqqara at least three-quarters of a millennium earlier.

Almost every reference in the Bible shows that the swarming locust was regarded as a visitation of Divine wrath. Thus, *Exodus* and *Joel* have their counterpart in *Revelations*: 'And there came out of the smoke locusts upon the earth; and unto them were given power, as the scorpions of the earth had power.' And in *Chronicles II*: 'If I shut up heaven that there be no rain, or if I command the locusts to devour the land . . .' The assumption in all cases is that the sinful must be bitterly aware through previous experiences of the nature of the threatened punishment. Habituation to the locust and close observation of its appearance and ways are also shown in a number of passages still remarkable for their exactitude.

'The locusts have no king, yet they go forth all of them by bands.' The reference, in *Proverbs*, is to the hopper bands which precede swarm formation. 'And the shapes of the locusts were like unto horses prepared unto battle; and on their heads were as it

were crowns like gold, and their faces were as the faces of men.'
This, from *Revelations*, is echoed in *Joel*'s 'Their appearance is like
the appearance of horses, and like war horses so shall they run'.
Again in *Joel*, there is '. . . a nation has come up against my land,
powerful and without number; its teeth are lions' teeth, and it has
the fangs of a lioness'.

All of these descriptions would be approved by an entomologist.
Looked at from underneath, the mouth parts of *Schistocerca
gregaria*, the Desert Locust, do bear a strong resemblance to a
lion's muzzle surmounted by nose and eyes, so much so that
Egyptian faience workers embodied this aspect especially in their
representation of the Desert Locust. The upper part of the head,
when viewed from the same angle, looks, moreover, very like a
crown and the gold is the metallic yellow of mature insects in
their swarming phase. The muzzle can also be viewed, without
overstretching the imagination, as a human face, while the head,
seen sideways on, could be that of a stylized horse. German
peasants, noting a similar resemblance in the grasshopper, called
it a Hay Horse – *Heupferd*.

The locusts' seasonal movements were also recorded in the long-
distant past, so that *Exodus* speaks of the east wind that brought
the locusts into Egypt and the west wind that carried them away.

In later pages I shall deal at length with modern researches into
the connection between meteorology and the build-up and flights
of swarms and need only to remark meantime that an Egyptian
entomologist, Abdel Megid Mistikawy, noted some three and a
half thousand years later (in 1929) that 'All the worst Egyptian
invasions have come from the east on an east wind'.

To assess the havoc wreaked by the Desert Locust during
thousands of years of infestations is impossible, partly because early
writers made no distinction between deaths caused by famine and
others resulting from pestilence, which they often associated with
locust plagues, but also because, until comparatively recent times,
the various locust and grasshopper species were constantly being
confused. This is not surprising. A locust is a locust to the farmer
who suffers from it.

One of the first to describe a particular plague was St Augustine,
who in *The City of God* recalled that 'when Africa was a Roman

B

province it was attacked by an immense number of locusts. Having eaten everything, leaves and fruits, a huge and formidable swarm of them were drowned in the sea. Thrown up dead upon the coast, the putrefaction of these insects so infected the air as to cause a pestilence so horrible that in the Kingdom of Masinissa alone 800,000 and more are said to have perished. Of 300,000 soldiers in Utique, only ten remained . . .' Stories of drowning locusts have often been confirmed, most frequently around the Red Sea, where reliable witnesses have reported seeing their bodies piled three or four feet high along the coasts for several miles. Even though St Augustine's figures of the dead populace seem sensationally high, they are not impossible or even out of line with some modern famine figures. The cause of death was probably hunger followed by typhus, consequent upon a total loss of the country's crops. Other writers, describing other plagues, have said that the dead locusts caused an epidemic among animals feeding on them and that many men died of eating infected meat. Reports of deaths caused by putrefaction of locust bodies have also come from Russia, Poland and Lithuania. In A.D. 591 a migration of locusts from Africa into Italy is said to have caused a pestilence carrying off nearly a million men and beasts

This ancient belief in the power of the locusts to bring with them an epidemic disease of one sort or another was obviously fostered by the sight of huge piles of their rotting bodies when the plague died out. Streams and wells close by would be contaminated so that widespread sickness inevitably followed. From this observation of water-borne epidemics to the conclusion that the presence of locusts was in some way connected with all subsequent epidemics would be a natural step.

Detailed accounts of a disastrous European plague appeared in French journals in 1613. Significantly, they begin with an account of an intensification of the bad weather which had characterized the whole of the preceding year. Beginning in the Midi, as the year ended, terrible storms swept the European coasts, destroying ships and boats by the hundred. Along the French and English coasts alone more than 2,000 bodies were washed up, and 1,200 were recovered along the shores of Holland and the Texel narrows. In the port of Amsterdam ships newly arrived from the

East Indies were sunk before they could unload. Sixty-four ships of various nationalities were wrecked off Portugal by tremendous winds and in Italy most of the rivers, including the Po and Tiber, overflowed. Earth tremors struck Germany, particularly West-phalia, where the wreckage included much of the town of Bielefeld and part of the castle of Spangenberg.

This was followed in May of the same year by a plague of locusts, beginning in the Camargue, where they were described as looking like little grasshoppers. At first white, they became black within three days and greyish within the space of a month; they had 'teeth', double wings, six legs, three on each side, the back pair bigger than those in front. Fully grown, they were said to be about the length of a couple of thumbs. Local flights were of a distance of a league or more.

Within a few days they had made their presence known by their havoc as they gnawed the produce of field and garden down to the roots. Meadow grass sufficient to feed three or four thousand head of cattle was devoured in its entirety in seven or eight hours. Having finished off grass and other greenstuff, the swarms fell on the cornfields just as they were due to be harvested.

Having eaten out the Camargue, they passed on up the Rhône valley to Tarascon and beyond. There the corn had fortunately been cut, but not the lucerne, vital horse fodder, which they destroyed, together with every garden vegetable. Winter caught them in Burgundy and thanks to low temperatures and attacks by starlings and hordes of other birds they there died off.

Meantime, however, they had bred, laying their eggs at the end of the autumn in sand patches in what was described as a kind of tube enclosed against damage by weather. When these were dis-covered the magistrates of Arles, Tarascon and Beaucaire issued orders that they were to be dug up and brought in, presumably as a means of checking on their numbers before destroying them. Many more remained in the fields, where as many as possible were flattened with spades and sandbags by the anxious peasants and their families. Even so, hoppers seem to have hatched out all over the district, threatening the new season's crops; but fortunately the country people were now becoming familiar with their ways and particularly with the insect's need of shelter against the cold night air. Large numbers were caught by stretching linen sheets like nets near their natural ground cover, while others were trapped in banks of shrubs laid out as lures. By these and other means the swarms were destroyed before midsummer.

The fact that this plague followed on a period of Mediterranean storms of a kind and intensity now thought to be an important cause of some very big locust movements was not noted at the time. It would not have occurred to anyone then that two events apparently separated by four or five months could still contain cause and effect. No one, moreover, seems to have observed the arrival of the locusts or, even afterwards, to have conjectured where they came from. Nor is this surprising in an age when insect and animal plagues of all sorts, from butterflies to spiders, cock-chafers, flies, snails, slugs and frogs, were commonly believed to be the direct result of putrefaction, and even of the corruption of the air itself in hot damp weather – a belief still held in parts of Africa. These, of course, were also the atmospheric conditions most favourable to the development of a bubonic epidemic, and the two plagues, that of the locusts and that of the disease, were as a result closely linked in people's minds.

Looking at the evidence now, one can make a fair guess that the build-up for this invasion over three and a half centuries ago had actually begun not in May 1613, as the peasants and their magistrates imagined, but at some time in the previous autumn, when similar weather was being reported. Breeding unobserved in the comparatively unpopulated Camargue, they were then able to emerge in overwhelming numbers. These are indeed precisely the types of conditions and circumstances in which the Desert Locusts are still able to catch their victims napping. But in the seventeenth century none of this was known, nor were the descendants of the Tarascon peasants any better prepared when, in 1720, another locust plague struck them.

This also occurred in May and was so disastrous that on the thirtieth of that month an ordinance was issued on behalf of the Archbishop of Avignon, of whose diocese Tarascon was part, instructing the priests to add a prayer for the cessation of the locust plague to all masses and benedictions during four days. The prayer was reinforced by a ritual of exorcism in which all the religious corporations and town and district countryfolk were required to join. Meeting after vespers at the church of St Marthe, the procession made its way to all parts of the countryside which seemed to be suffering worst. M. Declerk, the official over whose name the ordinance was issued, explained that the Archbishop would himself have performed the ceremony but for his illness, and exhorted all to pray for the recovery of his health, 'so necessary to all the diocese'.

To the first part of the ordinance, which was later bound and filed with several others dealing with the terrible outbreak of bubonic plague beginning soon afterwards, the Archbishop's secretary later added an asterisk against the word '*sauterelles*' accompanied by a curious marginal note: '*Locustae pestis nunciae*' – Locusts, harbingers of the plague. In fact, within a week after the issue of the ordinance the ship *Le Grand St Antoine* had tied up in the port of Marseille, carrying bubonic germs, and although all her passengers and merchandise were immediately put into quarantine in the infirmary outside the town it was already too late: within a short while the epidemic had broached the walls and was raging through the city. It continued for three years, ravaging the whole of Provence. To peasant and primate equally, it seemed that the locusts must have brought it. What they had done in reality was so to weaken the people by starvation that the bubonic plague was able to run unchecked.

There are many similar examples of the belief that locusts were the cause of sickness. In Florence and elsewhere in Italy in 1478, for example, an outbreak of pestilence was said to have been preceded by a locust plague. There are also many Biblical references in which the association of the two kinds of plague is constantly made.

Most of these accounts have been taken from countries where outbreaks were comparatively rare and, therefore, seemed more noteworthy. They were seldom long-lasting and were probably caused by various locust and grasshopper species. In the invasion area of the Desert Locust, the infliction is the greater because the peasants are the poorer and their livelihoods are at the best marginal. The land is poor, the work hard and everything grown by hand can be destroyed in a few hours. Travelling through the deserts one is struck again and again by the vulnerability of the

oases and even of the more extended crop areas like those of the
Nile valley. One way and another this huge region, more than
twice the size of China, supports well over 300 million people to
whom the locust presents a perennial threat. Very often it seems
as though it is the worst troubled areas which are most in danger.
In 1889, when the British were making war in the Sudan, a plague
of classic proportions fell on the fields beside the Nile with such
devastating effect that when the locusts passed on there was
scarcely enough grain left for even the mice to scavenge. These
mice, too, assumed plague dimensions, so that for the inhabitants
of Omdurman, crowded in their hovels, nothing was left. Some
turned cannibal; for others the end came in death on the streets
or the banks of the river, whose waters were soon black with their
putrefying bodies floating downstream to spread disease in the
locusts' wake.

In the circumstances of disaster like this it is impossible to
separate the degrees or kinds of misery which ensue. Even today,
although it is easier to rush aid to a stricken community, the cost
of control in terms of labour and other services which have to be
re-directed from other uses is grievously heavy for the countries
under attack. The figure of about $100 million a year suggested
for the fourteen-year plague which ended in 1963 takes no account
of such direct and indirect losses as the abandonment of cultivations and the disruption of trade. In India, for example, the value
of the crops lost to the Desert Locust between 1926 and 1931 was
reckoned to be equivalent to £4 million, but the consequent
losses due to premature death of livestock for which there was no
feeding stuff proved incalculable. In Ethiopia in 1958 cereals
sufficient to feed a million people for a year were eaten by locusts
instead and although with characteristic generosity the Americans
were quick off the mark with famine relief in the form of 20,000
tons of imported grain it was still necessary for the Ethiopian
government to make large-scale tax remissions to the afflicted
farmers with consequent set-backs to the country's whole economy.

In the winter of 1954-5, when vast swarms of locusts arrived
from the South Sahara to hole up in the Souss valley of the
Moroccan Atlas mountains, there were scenes which must seem
incredible to those who have never seen a large-scale infestation.

This is an important citrus fruit-growing region and the value of crops in this valley alone was estimated to be £4½ million. In one respect the situation was ideal for a large-scale counter-attack by poison-spray planes, for the cold of the enclosing mountain tops prevented the insects from moving on. Unfortunately, their sheer weight was their own best defence until the damage had been done. Carpeting roads and orchards a foot or more deep, they wreaked total destruction. When the pilots of the control planes had done their work, the burning of the locusts' bodies began; and when the smoke cleared it was seen that not a tree had escaped, and scarcely a fruit remained to be plucked.

The frequency of locust plagues and the fears they cause have given rise to other, supernatural, fears, the most extreme being that if the locusts *fail* to come another even worse fate may befall. Kamal-ad-Din ad-Damiri, a noted Arab zoologist of the fourteenth century, recorded that the Kalif Umar Ibn El Khattab, having seen no locusts one year, became so uneasy that he sent out three mounted messengers, one to Yemen, one to Syria, and one to Iran, to inquire if any people there had news of them. Not surprisingly, it was the messenger to Yemen – now known to be a high-incidence outbreak area – who brought back a handful of the insects. When they were spread before him, Kamal-ad-Din related, the Kalif exclaimed with joy, for, he said, 'I have heard the Apostle of God say that God has created a thousand nations of creatures, six hundred of them in the sea and four hundred on land, and that the first one to perish out of them will be locusts, which, when they perish, will be followed by other nations like a string of strung pearls when the string is cut'. Another legend, believed by the same Kalif, was that the Prophet had himself received a warning of a quite different kind when a locust fell before him. On its wings he found written in Hebrew: 'We are the army of the Great God, and we lay 99 eggs; if the number of these is completed to 100 we shall eat the world and what there is in it'. The Desert Locust does, in fact, commonly lay upwards of 100 eggs.

But if the locust eats out man, it is also true that over wide areas of the world where plagues habitually occur man takes his own small revenge, and makes good a minimal part of his protein losses, by himself eating the locust. In Islamic countries this has

led to some fine theological hair-splitting ranging round such
questions as whether the locust is to be regarded as a creature of
sea or land. There was, and is, a natural confusion about this, for
when swarms are seen coming in over the sea, as happens often
in the Indian Ocean, the Red Sea and the Persian Gulf, it seems
evident to simple people that they must originate there. Kamal-
ad-Din said that the Prophet himself held that two dead animals
and two bloods, fish and locusts, liver and spleen, were lawful
eating and quoted evidence that his wives had sent presents of
locusts on trays to him and his followers, one of whom observed:
'I wish I had a basketful.' There was also, however, the question
of *how* they were killed. Four imams stated that the eating of
locusts is equally permissible whether they have died a natural
death or have been caught and killed by a Magian or a Muslim.
Whether any part of the locust had been cut off, they added in
response to another inquiry, was irrelevant. On the other hand, it
was said in the name of Ahmed that if they had died from cold
they ought not to be eaten, while yet another school held that if
their heads were cut off they were lawful, but otherwise unlawful.

After all this, one may ask what locusts taste like. Abu Osuran,
another Arab zoologist, compared them when roasted to roasted
scorpions and said that they both resembled chicken meat. Others,
comparing them with prawns, have declared them to be most
delicate when the female is gravid.

F. S. Bodenheimer, in an entertaining book, *Insects as Human
Food*, traced the eating of locusts back to the eighth century B.C.
and the time of Nineveh, where, in a nearby palace, locusts are
depicted in a relief being carried on skewers to a royal banquet.
This is thought to have been the Desert Locust, the same which,
according to some translations of *Leviticus*, is stated to be lawful.
St John, of course, nourished himself in the wilderness on locusts
and wild honey. To Greek historians locust eaters were known as
acridophagi. Diodorus of Sicily described those of Ethiopia as small,
lean, spare and exceptionally black men who, when the south
winds rose in spring, lay heaps of wood throughout a valley along
which the locusts were expected to pass. When the swarms arrived,
driven northward on the monsoon, the *acridophagi* lit the fires,
suffocating and 'blinding' the locusts so that they fell to the ground

and could be gathered into heaps. Well salted, they provided food for the year round. The people, as a result of this diet, Diodorus reported, were short-lived, seldom exceeding forty years. He believed that winged lice bred in their flesh, first in their breasts and abdomens, then quickly consuming the rest of the body.

This early, general description of a swarm movement has been confirmed by repeated modern observations. The valley in which the Ethiopian *acridophagi* laid their fires to trap the locusts must have been one associated with the Great Rift through which spring generations of the Desert Locust often migrate northward from Kenya, Uganda or even Tanzania during plague periods. The experienced eye of the peasant would have noted that this occurred at fairly specific times, depending on the winds and rains, and would be ready for them.

But to return to locusts as food. The method of preparation among dwellers on the edge of the Grand Syrte, Herodotus observed, was to dry the insects in the sun, then pound them to a powder to be mixed with milk. Among the Attic shepherds, Aristophanes declared, locusts were much enjoyed. Roasted and dressed with vinegar and pepper, they were exposed for sale, he said, by poulterers. Elsewhere in Europe, the locust has been eaten from Russia to Germany, where it was served at the Council of Frankfurt following a great plague in the country. Diets of insects are common in parts of Africa, where the locust has naturally formed part of them.

Livingstone, who must have had experience of at least four species, regarded them as a blessing to the poor, who pounded them to flour, which, when well salted, kept for months. He disliked them boiled but found them of a palatable 'vegetable' flavour when roasted and preferred them on the whole to shrimps. He also compared their flavour to that of caviar. R. Moffat, a Scottish missionary who wrote an account of his African adventures in the mid-eighties, likewise regarded locusts as being a benefit rather than otherwise. But this was in southern Africa, where the Desert Locust does not penetrate, and *his* natives prepared them by steaming, afterwards spreading them out to dry in the sun on mats. They then winnowed them of their wings and legs and put them aside in sacks or heaps to be eaten whole as required or made into flour and mixed with water. He too thought they were almost as good as shrimps.

The Arab preference – and who should know better, for it is Arab countries which have been among the chief sufferers – is for grilling, roasting or, in Morocco, smoking. They may also be eaten boiled with couscous or fried alive in boiling oil, when they are said to have the flavour and consistency of the yolk of hard-boiled eggs. In Arabia the flour of locusts is sometimes made into cakes and eaten as bread. Doughty, however, said that some nomads were put out of countenance when they admitted that they ate so wretched a food; and it is indeed as a food of necessity rather than of choice that most people would consider it, whether in this or any other form. The very anxiety with which Muslims have regarded it is an expression of their doubts; and those doubts must spring from revulsion to what it symbolizes: destruction, poverty and death.

There is one other aspect of the locust in social history which is worth mentioning, for it leads us on to the whole subject of the battle now being fought against it. This is the variety of methods used in an endeavour to check depredation. Like so much else concerned with this mysterious insect, they run the gamut from superstition to practical trial-and-error. Some examples of the more practical measures have been given above. According to Palladius the ancient Greeks were at the other extreme of thought and held that the best way to ward off an infestation was to retire indoors as soon as the swarm appeared, when they would pass over without doing harm. If this ostrich-like method failed, the best way to get rid of them was to sprinkle them with an infusion of boiled bitter lupins and wild cucumbers mixed with brine. (First catch your locust.) Another way was to burn some as a warning to the others. Pliny, who gave some very exact descriptions of the scale of locust migrations, including those over the sea, as well as of their ability to withstand prolonged periods of hunger when necessary, said that in Cyrenaica it was obligatory for the citizens to make war on the locust three times a year, destroying first the eggs, then the hoppers and finally the mature insects. All who failed to do so were punished as though they were military deserters. The Syrians, accustomed to large-scale locust invasions, used their army against them, as have many other nations since, notably the British during the Second World War.

Inevitably, all the older methods were crude. It was not until 1885 that poison baiting with sodium arsenate was used against the locust, thus paving the way for the advanced chemical techniques of today; but this method had to wait until the twentieth century before it became widespread. Before then, and for long afterwards, the farmer's chief weapon against the oldest of his adversaries was his muscle as he dug defensive trenches in the baking soil and did his best to drive the seething creatures into them to burn and bury. This method could be used chiefly against hoppers and was not particularly effective. But it was the only way he knew. When the country's authorities conscripted him into the work it meant, moreover, that he usually received some poor payment in return for the loss of his livelihood, which was likely to occur anyway. So important was this that many men, finding themselves superseded by new techniques, felt themselves to be the victims of yet another blow by a malicious fate. This was not the least of the cruel ironies that have attended the history of *Schistocerca gregaria* in proving itself to be one of the most tenacious and dangerous of creatures next to man himself.

Chapter 2

An International Enemy

The history of international attempts to control the most highly nomadic pest in the world began, like most international hopes, with plenty of good will and no results. Invited to a conference convened in Rome in 1920 by the Institute of International Agriculture (forerunner of FAO), fifteen nations sent delegates, ten signed the ensuing convention, the rest (including Britain) went placidly home, leaving the matter to God, Allah, the suffering farmer and any small mercies his individual government could provide. In 1928 another plague occurred and this time there could be no complacency. With great swarms invading Anglo-Egyptian Sudan and Kenya and some getting as far south as present-day Tanzania, an alarmed British Colonial Office asked for a committee on locust control to be set up to advise what should be done. This was important, for by far the greater part of the Desert Locust's invasion territory lay in British colonies or associated lands, and it was this British action, undertaken with a characteristic mixture of enthusiasm and cheeseparing economy and practically no staff, that was to lead, indirectly, to the full international collaboration that we have now. It was also to result in the biggest bargain in entomological history. A few years earlier a small untiring man of original genius, the Russian entomologist B. P. Uvarov, who had done much work on locusts, had arrived in England as an *emigré*. He was therefore the natural choice to organize the necessary investigations; he was already on the staff of the Imperial Bureau (now the Commonwealth Institute) of Entomology, which continued to pay his salary; a quasi-governmental body, the Empire Marketing Board, was willing to put up the rest of the money; and so for an all-in annual cost of £422 4s. 3d. the locust-afflicted colonies got the services of a headquarters organization consisting of two people, one being the world's leading authority on the subject, B. P. Uvarov, and the other a young girl, Miss Zena Waloff, who was herself to become a major expert. I shall discuss their work in detail later; sufficient to say now that it was the systematic identification of species and charting of reported locust sightings, begun in those early days, that laid the foundations for the strategy of control which is now in operation wherever the locust flies.

Uvarov had two good allies in an old friend, Sir Guy Marshall, director of the Institute, and Francis Hemming, a British civil servant who was himself a fine entomologist and acted as secretary of the locust control committee. With their help in opening doors, this somewhat amateur British conception of how to tackle a major problem without actually spending money was probably the best approach at the time, for, not being answerable to the British Treasury and with no political strings attached, Uvarov's little organization, working as part of the Institute, was able to cut corners and get to the heart of the matter without interference. From the first, Uvarov was convinced that in dealing with an insect able to cross so many frontiers so easily it was useless to confine studies to only a section of its invasion area. In order to build up an accurate record, therefore, he wrote to the governments of the affected countries, whether colonial or foreign, asking for their aid in keeping watch; circulated illustrated pamphlets describing what they should look for to consuls, missionaries, military men, oil explorers and others whose duties might take them into the deserts; and, in short, badgered everybody likely to be able to help. The resulting flow of information, centralized through his office, made possible what was in practice a Desert Locust intelligence service. This was not only of aid to the countries afflicted. It was the first example of international co-operation in action.

In 1935 and for some years afterwards the Desert Locusts again

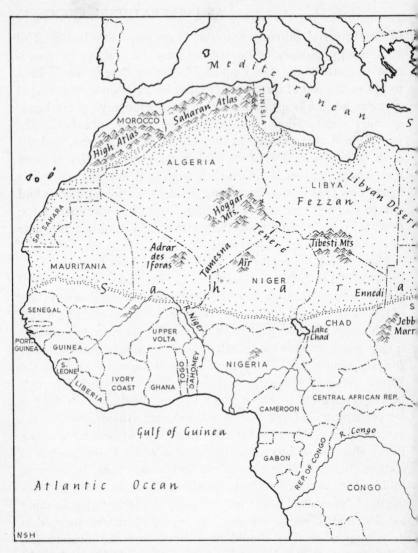

1 Topography and the Desert Locust. In order to breed successfully the Desert Locust requires enough rain to moisten the sands for egg-laying and encourage greenery for food and shelter. These conditions are most likely to originate in the proximity of mountains, rocky massifs and certain dunelands. Although heavy rain can run off for hundreds of miles through normally dry

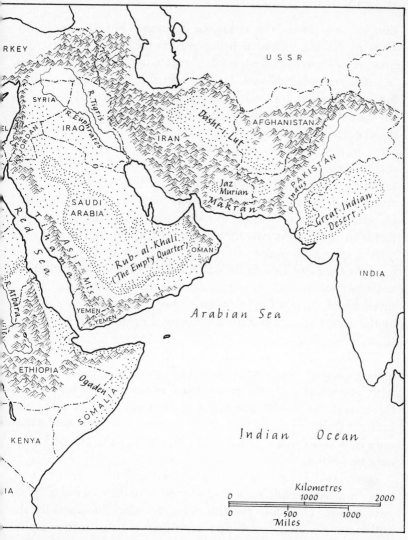

wadi beds, or in the form of flash-floods, it should be possible, with improved meteorological information, to narrow down the areas needing to be surveyed and searched in order to control outbreaks. But the total area will always be vast and international aid and cooperation will always be necessary.

disappeared into their unknown fastnesses. In the meantime Uvarov, having gained a little more money, sent the first of a number of experienced entomologists into the field to, as he put it, 'wander around and look'. He was also able to set up a one-man locust laboratory for such fundamental and practical studies as the influence of temperatures and humidities on the breeding and development of locusts and the ways in which contact pesticides work in destroying them. (It was at first thought, after experimental dust-spraying of swarms from the air, that the insects were killed as a result of inhaling the poison, but this proved to be wrong and the money ran out before alternative theories could be tested. Chronic shortage of money was in fact the bedevilling factor in much that was then being attempted and it was only after a successful plea to contributing governments and a grant from the Carnegie Corporation in America, then slowly awakening to its own locust and grasshopper problems, that the situation was saved and Uvarov's organization – now known generally abroad as the International Centre for Locust Research – able to continue.)

The next and, in our story, one of the most important and longest plagues, was a wartime one. Beginning in 1941, it lasted, with one short lull, for more than twenty years, for a time threatened the Allied war effort, ruined countless farmers and brought death to many and made it clear beyond a doubt that only by a concerted international effort would the Desert Locust ever be beaten.

'War,' said Uvarov when I was discussing these matters with him, 'is a wonderful factor for progress – at least technically.' It gave him his chance to get full and direct official support for the anti-locust research centre (it was still not entitled to the capital initials of the ensuing body of the same name), not to mention undreamed-of quantities of field personnel when the crisis broke. The first indications of the new plague came, in the spring of 1940, from India, where important investigations into the species had for several years been directed by a brilliant entomologist, Rao Bahadur Y Ramchandra Rao. The reports by the Indian Locust Warning Service spoke of many locusts, not yet swarming, invading from the west. This was followed, during the 1940 monsoon

1*a* Locusts from a bas-relief frieze at Saqqara, Upper Egypt, about 2400 B.C.

1*b* Locusts on skewers as an offering to the table of the Assyrian king Sennacherib. Bas-relief from Nineveh, *c.* 8th century B.C.

2 The six stages in the life-cycle of the Desert Locust (*Schistocerca gregaria*). Left to right, the first five insects represent successive instars, stages between moults before flying begins; the sixth is a young adult, which can fly great distances and might be one of a swarm numbering several hundred millions. In all, the Desert Locust may live four to six months, the female laying three to four hundred eggs in various locations. The adult's wing-span is approximately four inches (10 cm.).

season, by breeding in north-western India, where swarms began to form. Some moved out to start off a fresh round of breeding in Persia and Arabia, launching yet more swarms, with the result that by the summer of 1941 the plague reached Egypt, Anglo-Egyptian Sudan, and Eritrea, whence, within a matter of weeks, more swarms moved into British Somaliland, Somalia and eastern Ethiopia, threatening the whole of East Africa. In the current war situation this was extremely serious, for many of the imperilled districts were close to Allied lines and, being politically unstable, they only needed famine to produce potentially disastrous conse-quences both for themselves and the Allied armies. Yet with shipping desperately needed elsewhere, it was impossible to divert food transport. All that could be done was to treat the locust hordes as more than normally dangerous enemies and endeavour to counter-attack accordingly. A full description of the ensuing campaign, the most arduous ever fought in the long history of war against the pest, has never been published. Uvarov's own brief and modest account, plus a few notes on one of its principals, will therefore have to do.

Although the campaign was mounted by the British, it was international in that at one time or another every Ally and friendly country in the invasion area contributed to it ('sometimes,' said Uvarov wryly, 'by not preventing us'). It was, moreover, the first time that a single campaign plan had been effectively applied over practically the whole of the Desert Locust's natural region. Only the French West African colonies were, for military reasons, omitted. In England an Interdepartmental Committee on Locust Control was established in order to ensure that every government department could be called upon to help, as required. The co-operation of the armed services became a top priority. The strate-gic planning was to all intents and purposes entrusted to Uvarov and his now-strengthened centre.

The ten years spent in compiling charts and data concerning locust breeding and migrations while the centre was struggling to keep alive now showed their value, for the work which was then done pointed conclusively to the Arabian peninsula as the key area for the main effort. Three exploratory missions sent to the north-west, west and south-east gave a good idea of the problems which

C

would have to be overcome. One of these parties, in fact, performed the feat, first achieved by Lawrence of Arabia in the reverse direction, of a camelback crossing of the Great Nefud desert. The reports they brought back showed that a motorised campaign, although difficult, would be possible, and a para-military organization was accordingly set up, with Uvarov and his colleagues providing the technical direction. The resulting expedition in 1943-4 was one of the strangest of the war. In two vast convoys 24 officers and 803 other ranks, all unarmed at the insistence of the Saudi Arabian government, and with attendant wireless and medical sections, set off from Cairo in some 360 vehicles weighed down with loads of sodium arsenate for the locusts and bags of specially minted golden sovereigns and silver Maria Theresa dollars with which to pay for recruited labour. Among the civilian personnel were half a dozen locust officers, including one American and a Sudanese, and seventy Sudanese supervisors who had been trained by the Sudan locust organization. Both convoys were to set up supply bases from which field operations against the locusts would be carried out. One crossed the Sinai desert and made for the Hedjaz and Yenbo on the Saudi Red Sea coast; the other had the enormous task of getting itself across 2,500 miles, mostly desert and very largely roadless, through Baghdad, Basrah and Kuwait to Dharan on the Persian Gulf. 'Difficulties experienced and overcome, during these approach marches, as well as during the subsequent operations, were numberless,' wrote Uvarov. 'Routes had to be found across belts of sand-dunes, lava fields, or dry river beds liable to become raging torrents during a rain and treacherous morasses after it. Scarcity of water, burning heat in the daytime and bitter cold at night, mosquitoes and fever in coastal areas, monotonous rations and lack of contact with the outside world' – all were discomforts cheerfully borne by soldiers and civilians alike.

In Persia it was also necessary to use troops. They were led in the technical operations by a British entomologist famous in locust work, O. B. Lean, who in the years before the war had survived terrible experiences, virtually wrecking his health, in tracking down the breeding places of the African Migratory Locust in the swamps of the Middle Niger during the rainy season. Speaking about him, Uvarov told me: 'Because of his ill health he had taken a less exacting job in England. Then, when the new outbreak began, he came to my office and said, "Oh, if I could only go somewhere!" So I said, "Well you can – you go to Persia!" Within a week he was off and within a fortnight of our making the first plans for the campaign he was at work.' Surviving a plane crash soon after his arrival, Lean welded together a unique organization of men, most of whom knew nothing of locusts but were willing to learn from the experts, who included Indian entomologists trained at the famous Lyallpur Agricultural College. In India, where anti-locust control had been split up between the provinces, a coordinated scheme for the whole of India, including present-day Pakistan, came into being. In cooperation with Lean's little army it was thus possible to protect the vital border provinces. While the Anglo-Indo-Persian effort enabled the main military supply route to Russia to be protected, Russian experts were also able to help the Persians. The combined anti-locust force in this region during the last year of the war included Persian, Indian, Russian and British experts and technicians, a unit of the Indian Army and a Russian air unit of ten aircraft. Brought together in anticipation of an invasion of Desert Locusts from India (which did not, in fact, materialize), they represented international cooperation at its peak.

In East Africa the plague was at its worst in Sudan and Kenya, where at one time 4,000 troops and 30,000 civilians were in action, hacking roads through the bush so that they could get to the infested districts. The total cost of the Sudan campaigns was nearly £1 million in six years; in Eritrea a year's campaign alone cost £150,000. Were there red ears, one wonders, in the circles of British officialdom which had grudged Uvarov his £500 a year for a control and research centre? If so, he bore no malice but devoted himself to creating a post-war regional organization for East Africa and the Middle East which would be able to carry on when the impetus of war ran out.

At the end of 1947, when a lull occurred, it was possible to assess the results of the first six years' campaigns. Only in Sudan and Eritrea had there been substantial losses. Elsewhere the crops had been effectively protected. New techniques of control had

been developed, a new and valuable pesticide – BHC (benzene
hexachloride) – had been introduced. Lessons were being learned
continuously over the whole period. One was that a planned over-
all offensive against the Desert Locust was a practical and worth-
while proposition. Another was that this locust presents a litmus
paper test of man's ability to cooperate; and in 1947 the signs of
his doing so were not particularly encouraging. The wartime
conditions that persuaded nations to drop their political enmities
had been exchanged for peacetime wrangles; another plague, or a
revival of the same one, was needed to bring about a change of
heart. When it did so the lead in the campaign could no longer
be taken by Britain, though her scientists, Uvarov included, continued to play a vital part.

The whole attitude of the 'have' to the 'have-not' nations had
also undergone a change. In 1945 the Food and Agriculture
Organization had been brought into existence as the first of the
United Nations agencies charged with the task of aiding the
'have-nots' in development. There were also the first signs of the
population explosion which was soon to present the world with
major problems. These two factors – development and the growth
of population – meant that more land must be cultivated, using
improved methods. And both of them favoured the locust. As
early as 1936, at a meeting in Cairo, Uvarov had pointed out that,
contrary to the common belief, shared even by some entomologists,
outbreaks of many species of locusts and grasshopper become not
less severe but more so as development increases. Their excessive
multiplication and spread and the formation of areas suitable to
outbreaks are furthered, not hindered, by extended and improved
cultivation. Examples given by Uvarov included the clearing of
forest in favour of grasslands which had encouraged the Oriental
Migratory Locust in Borneo and Malaya, where the breeding of
swarms would otherwise not have been possible; the overgrazing
of mountains bordering the southern Mediterranean fostering the
Moroccan Locust in Cyprus and elsewhere. The connection between the Red Sea coasts and the Desert Locust's disposition to
break out there might be partly explained, he thought, by the
system of native cultivation dependent on seasonal rains and
floods. His most striking example of grasshopper plagues on a

serious economic scale, however, came from no poor country but from North America, where they had clearly followed agricultural development.

The thought that one aspect of its policy of development might be endangered by another had occurred quite early to FAO, but the organization was functioning without sufficient funds and there was little it could do when renewed outbreaks began to be reported, at first from Arabia, where cyclonic rains had softened the sands of the Empty Quarter in October 1948, creating green habitats for the Desert Locust where earlier there had been nothing, and, a year later, from India, Pakistan and Iran, whither the swarms had migrated. A plea from India on behalf of herself and her neighbours had to be turned down, and, although detailed campaign plans were eventually drawn up following a conference convened by FAO in Rome, they could not be put into effect. For want of a relatively trivial sum (about $1½ million) the chance of averting the disasters of the next ten years was lost.

Another temporary casualty caused by the renewal of the plague was Uvarov's dream of a thorough-going field research programme for studying the ecology of outbreak areas in eastern Africa and Arabia and the conditions which cause innocent solitary insects to become gregarious and swarm. Some important research was in fact carried out, but the organization set up to do it, the Desert Locust Survey, quickly found itself swamped by control work against the vast numbers of locusts breaking out on both sides of the Red Sea and countries to the south. With the British Empire breaking up, it is rather remarkable that the British government of the time maintained locust work at all. Creditably, they did so, and Uvarov had the joy of seeing the Anti-Locust Research Centre, the baby he had nursed so long, become a government-sponsored body with both status and an assured budget. Twenty-five years later, it remains the most important auxiliary within the international effort. In 1952, when FAO, having found some money, was getting into a first-gear start with its first international Desert Locust project, the advice of the British centre was vital.

The history of the international campaigns of the next ten years can be summed up by saying that they saw, for the first time, a gradual, sometimes grudging, often capricious, but on the whole general acceptance of the fact that while each nation retained overall responsibility for controlling any locusts within its borders it could not do so successfully alone, but would need to pool its information, expertise and technical resources as far as possible with its neighbours and co-victims; and that, moreover, this could only be done efficiently within the general strategy, which was now being supplied by FAO. One of the first tasks of the indomitable O. B. Lean, who coordinated the national campaigns on FAO's behalf, was to convince a number of the afflicted countries that the Organization was not a universal aunt with a long pocket for providing all the wherewithal. What it could and did do was to build up its own fleet of vehicles and machinery and establish strategic reserves of pesticides and equipment on which the various national anti-locust organizations were able to draw as required. It also bought its own aircraft – four Austers – with which to develop new techniques of air-spraying. For the rest, it was made clear that while FAO's main role was to advise and coordinate, that of the individual countries was to cooperate. This they did with some success, particularly in the plague's epicentre, Saudi Arabia, where at various times teams of seven different nationalities were operating.

Whether the plague died out in 1963 as a result of these and similar efforts or whether natural events – the failure of rains, or the lack of helpful winds – were the prime cause is uncertain. Certainly the last known swarms of this period were given their quietus by a hard-pressed campaign, including air-spraying, along the Pakistani-Indian border; but it would be a mistake to assume that this was any more than a telling blow. While the plague was in progress much had been learned, particularly about locust ecology; theories purporting to explain what happens during recessions and how plagues recur had been propounded, the riddle of migrations laid bare and important advances in methods of control had been made; but nobody had actually *seen* all the stages of an outbreak and until this was done, and the inter-relationships of insect, weather and habitats were fully understood, the prevention of outbreaks would be uncertain at best, and in most cases impossible.

Uvarov's cry – Research! Research! Research! – was never more needful of an answer than in the recession which followed. He was himself now in his still-vigorous seventies but while continuing to work at the Anti-Locust Research Centre he had relinquished its direction, content to know that new men and new methods were at work.

An important field research project was already in its first stage. Initiated in 1958 by FAO and UNESCO, it was an ecological survey of the kind Uvarov had always advocated, taking in all the desert regions and marginal lands where outbreaks of Desert Locusts seemed most likely to occur. In reality, this meant asking what happens to the Desert Locust *between* plagues. When and in what kind of physical surroundings does it survive, and under what conditions? This information is essential if one is to construct a picture in depth of the developments which might lead to an outbreak and to narrow down the localities in which searches should be made before swarming occurs. In the past, control work had always had to be done after the floodgates had burst, when the defenders had all their work cut out to avoid being overwhelmed. But if outbreaks could be anticipated the odds favouring the locust would be shortened and the plague curtailed or even prevented.

The ecological survey, led by George Popov, a member of the Anti-Locust Research Centre whose name will recur frequently in this narrative, was the most widely ranging geographical reconnaissance of its kind ever undertaken. Beginning with a six-month journey of 8,000 miles through Ethiopia, Sudan and Chad, it was followed by one of 15,500 miles across the countries of the southern Sahara, including some of the wildest regions of the desert itself, lasting a year. Although this first phase of the survey confirmed that the Desert Locust favoured certain kinds of habitat and had preferences for particular plants within it, this was by no means the whole story. Desert vegetation can be uniform, with a flood of plants of a single species filling a wadi* from end to end; it can form a mosaic, a patchwork of species; or in a comparatively small area it may vary from one extreme to the other with many gradations. The relationship between plant cover and the space of bare sand between plants is also significant, particularly when

* River bed, superficially dry except during floods.

the insects are ready to gregarize. Some of this was known or suspected before the survey but needed confirmation; a great deal more was not known at all. Like most fields of research when newly explored, the first phase of the ecological survey therefore revealed a gap. Filling it in required not three years, as originally planned, but seven. This extension was only made possible by a big injection of financial help by the United Nations Special Fund (now the UN Development Programme) enabling Popov to complete investigations in Africa and then move on to a 14,500-mile reconnaissance of Arabia, followed by thirteen months in Pakistan, India and Iran. His final score was 67,000 miles, all done by Landrover, truck or camel in some of the harshest country in the world. Two plant ecologists, Ch. Rossetti and W. Zeller, accompanied Popov at various times, but he alone of the experts went all the way. The team's findings provided for the first time a detailed picture of the needs and ephemeral opportunities by which viable remnants of the Desert Locust population are able to survive as individuals between one plague and another. They did not negate the theory that isolated swarms may also persist and help to carry over one plague to another, for there are, as we shall see, many factors in an outbreak. The immediate value of the discoveries made by Popov and his associates was that the search for the locust need no longer be a matter of chance but could be done systematically by ground and air teams who would in future know where and when to look.

Besides supporting Popov's extended survey, the UN Special Fund Desert Locust project, which came into existence in 1960, was designed to put campaigns against the locust on an altogether bigger footing, backed by massive research in field and laboratory, a strengthened and improved intelligence and forecasting service, the testing and teaching of new techniques of control and more and better training of national locust officers and technicians. The cost, about $4 million, was split in the rough proportions of two-thirds contributed by the Special Fund and the rest by some forty participating nations who were asked to pay according to their means and the degree to which they were liable to invasion and loss. About $325,000 went on vehicles and other equipment issued to key countries where it was especially important to look for locusts during recession periods, and the Desert Locust Information Service, already being operated by the British Anti-Locust Research Centre in London, was absorbed into the scheme and subsidized. FAO was entrusted with the running of the project, it being understood, as before, that its primary role was to co-ordinate campaigns and advise the affected governments, which would still have to find the personnel and do the field work. It would also be expected to contribute field officers and workers for special international surveys in suspected key outbreak areas, like Saudi Arabia and Iran, too large for the home government to tackle on its own. The operational research part of the project, on which hopes of developing new methods of control were pinned, petered out when the Desert Locust (which one sometimes suspects of having its own information service) went into recession soon after the project was launched. Although not all its immediate intentions were fulfilled, there is no doubt that the UN Special Fund, or, as I shall in future call it, the United Nations Development Programme Desert Locust Project, enormously improved the prospects for the future.

At FAO's Rome headquarters, meantime, two new faces had appeared. One was mine, as a consultant expert in the field of public information (a bureaucratic description of a news feature writer and reporter). The other belonged to an Indian entomologist and highly experienced Desert Locust expert, Gurdas Singh, who came from Addis Ababa to assume the over-all direction of the anti-locust campaigns. It was he who invited me to go out to the deserts to see the project at work and thus committed me to five years' unremitting involvement with an insect of which, until then, I knew nothing.

Chapter 3

The Life Cycle

All life in the desert is a battle. The rains come fleetingly, though sometimes with appalling intensity. Tamanrasset, in the Central Sahara, for instance, has an average rainfall of 1·6 inches a year. In 1950 nearly all of this was concentrated into forty savage minutes. On a date in December 1949, significant later as marking the onset of the longest of the great Desert Locust plagues, a five-year drought among the ash-coloured hills of the Red Sea coastal desert north of Jeddah was broken by a rainy season which included a deluge of seven inches in a night.

Pouring down from the mountains the accumulating water has the force of a moving wall. Under such tremendous pressure the primitive dams of the oasis farmer are broken down and his underground canals and their access pits choked and destroyed. Mud village walls literally dissolve. Boulders a yard thick can be shifted like pebbles. Each such torrent contributes a little toward the erosion of the landscape, aids the frosts to shatter it a little more and grinds the rocks and the gravels finer on the great plains endlessly stretching south. Yet these rains provide the wherewithal of any such life as the desert has, outside the oases. Without them any form of existence, even the most minimal, would be impossible. For as the floods surge down the valleys in temporary simulation of the rivers which once flowed regularly, they penetrate the ground, and then a miracle is performed. Where there was nothing but rocks, sand and gravel, a thin sheen of green appears. With immense speed it develops into a flush of herbage – grasses of several sorts, bushes which had withered to ground level, 'annual' plants whose seeds, in this country, may lie for years before they germinate, gourd-like fruits, hard and bitter, and flowers on thorny trees and shrubs. This will all be as ephemeral as it is sudden. The burning sun will wither it again. Within a month to six weeks there may be, at most, a thicket of pale brown stalks where the breast-high schouwia bloomed.

Within this time all the vegetation in a wadi many miles long must have performed its own act of adaptation to the hostile desert climate, growing, flowering and casting its seed ready for the next renewal. This may occur a year later, or may be delayed by six or seven years, and in this time the only growth will be that of the acacia trees and a few shrubs able to thrust their roots deep down to the lingering moisture. These, too, have adapted themselves to drought and heat. Instead of bearing leaves of the normal kind they have clothed themselves with thorns, reducing breathing to the minimum. Yet while the flush of vegetation continues in the wadi it will be a centre in which life is for the time being sustained. Any nomads in the area will find their way to the greenery to graze their goats and even their cattle on it. A camel owner will travel a hundred miles to pasture his herd there. Gazelles, which can race across the gravel plains at forty miles an hour, will come to an alert, uneasy halt. Lizards will appear; and there will be ants, gathering the seed of the tall grass, *panicum turgidum*, which the Tuareg will later recover from their ant-hills to eat as a substitute for millet. Small birds, arriving out of a seemingly empty sky, will also search the grasses for seed and insects; bigger ones, bustards, will hang overhead, ready to stoop on any prey that offers. And in this green mosaic the Desert Locust must shelter, feed and breed.

The cycle of life of *Schistocerca gregaria* is geared to survival in an environment of extreme hostility rather than, as might be thought, to plunder in an easy one. Thus the seething insect mass, able to ruin a farmer within a few hours, is only part of the locust picture, and is only really seen in plague times. The plagues themselves, as we shall see, are able to continue so long as a sequence of events and conditions remains favourable to the locust. When these go against it, when rains and vegetation fail at the times required for breeding, or for other reasons not yet fully understood its numbers drop below an unpredictable optimum, then it is no longer a marauder but a creature in retreat. There is no place in nature for it but the desert. This is not to imply that it falls back consciously or even instinctively to a kind of last redoubt in the central core of its vast invasion territories. It is rather that numerically weakened, the locust is contained in a series of them, widely separated and impermanent, where only its enormous adaptability enables it to survive until conditions are propitious for its next break-out. Until these arrive, everything is against it. Temperatures in the desert, as anyone sleeping out soon learns, can fluctuate by as much as 36 degrees C. (66 degrees F.) between day and night, and the same sands which have a summer surface temperature of 60 degrees C. (140 degrees F.) may crackle with frost in winter. The surface humidity of the soil runs to similar extremes depending on the highly erratic rains, and at neither end of the scale is it good for the locust. There is no single place in the whole of the desert breeding areas of over five million square miles where it may be sure of finding its needed habitat. This in a sense works to the insect's advantage in that man is unable to control it at such a point and may indeed even grow tired of the search; but it means that the Desert Locust must have extraordinary powers of flight to be able to traverse the vast distances required to reach where rains have fallen. And when this has been done successfully, there it may find its natural enemies also – birds, lizards, fly larvae and beetles, and almost all animals.

Basically, the Desert Locust is a winged big brother of its fellow-acridid, the familiar grasshopper of English meadows, and quite often leads much the same sort of life. Like other species of locusts, however, it has the peculiarity of being able to change its habits – to live two lives, as it were – and it is this characteristic that makes it so great a potential menace. The locusts I saw in the south of the Sahara or blundering into our tent walls on the high plains beyond Mecca certainly seemed harmless enough. But these were in their *solitaria* phase, isolated, almost lonely. It is when they veer round to the opposite pole of behaviour and begin to form into groups that they begin to loom with ill-omen. This is the gregarious phase, the *gregaria* of the insect's scientific name, and is, of course, the one familiar during the plague periods. The extremes of the phases are easy to recognize because they involve not only a changed way of life but changes of colour and even of physical measurements, but sometimes one comes across individuals of the species which appear to be poised uncertainly halfway between.

In the following section I deal with and describe an example of the Desert Locust, as it might be taken in the 'solitarious' phase fresh from a wadi.*

The first surprise is the hardness of the Desert Locust's skin. One expects something more pulpy, more obviously vulnerable. But this indeterminately coloured two-inch long body, something between brown and grey, or sometimes green, pinched firmly between the captor's thumb and forefinger, wears a suit of armour which is in fact both skin and skeleton. It is really a kind of box containing the creature's physical and nervous system and at the same time protecting it from the conditions it confronts. In the land of the locust everything is involved in a fight against desiccation caused by excessive heat and low humidity. The nomad tribesman has evolved his own form of adaptation by wearing loose robes and face coverings enveloping him in layers of air. The locust's box-like body-armour achieves the same result in a different way by incorporating a waxy outer layer which checks transpiration. Deprived of it, as the locust can be in a laboratory or in an exceptionally violent sandstorm, or at abnormally high temperatures, it becomes a brittle husk emptied of all moisture within a matter of hours. But this is rare.

Although there is a certain jewel-like charm in the tapering body shape and the acute angle of the hindlegs when folded as for basking, *Schistocerca*'s physical attributes are interesting rather than beautiful. Everything about him appears to be functionally designed without the power to delight. Although often confused

---

* 'Solitary' is regarded by entomologists nowadays as too ambiguous a description of a creature which in this phase, after all, may still be found in tens of thousands in a single stretch of wadi. I shall only use this word, therefore, when speaking of isolated individuals.

with the dragonfly, when seen at a distance, it is entirely lacking in that insect's lightsome grace. Its colour may vary in intensity with the freshness of the vegetation, and to the degree that it does so, will afford some protection. The head is broad and vizor-like with a cruel down-turned mouth and a pair of large protruding compound eyes capable of seeing from side to side through an angle of 220 degrees and up or down through 197 degrees. Such an immense range of vision makes the creature fairly difficult to catch. A shadow can cause alarm and we found it necessary to creep up from directly behind when possible, although a flying tackle, perfected by one of our Arab drivers, produced some fine catches at all angles. The value of such eyes, able to look above, below, before, behind, is, however, most evident in a swarm, when upwards of 120 million locusts in a square mile are keeping stations with each other, and their protective purpose is probably secondary.

Jointed antennae springing from the top of the head are used by the locust to make tactile inquiries into any other locusts it may meet and the ensuing fumbling and feeling has its own part to play in inducing a change from the solitarious condition to the gregarious one. The downward sloping mouth has been mentioned. Its hard black serrated jaws move sideways to cut through tough grass or the unfortunate farmer's wheat or barley, which is invariably severed just below the ear when it is on the point of ripening. Opening up our captive's abdomen, we should find some evidence of this voracious appetite, and the insect's need of it, in the form of a soft yellowish amorphous mass of fat stored after digestion of the food material. This is the insect's flight fuel and can keep it in the air for spells of up to seventeen hours in migrations well beyond the capacity of any other insect. Since even such feats of flying as this may still be insufficient to bring it to a land of plenty, nature has given it a safety factor, another result of aeons of adaptation, in the ability to slow up the rate at which the food passes through its gut. The forepart of this dark-coloured tube running from the mouth to the anus is, so to speak, the grinding and storage section, the middle is where digestion takes place and in the hind part water is absorbed. Although laboratory specimens have been seen to drink, those of the desert apparently do not and must rely for their water on the food they eat. When feeding is good the whole process, from the first bite to defecation of the little dark pellets scattered thickly around, normally takes from about thirty minutes to a couple of hours; but when food supplies are short and the Desert Locust's protruding eye, scanning the arid lonely land, finds no sign of green, then, in some mysterious way, the rate at which it converts its food will be retarded so that it may take from three to four days to pass through, and for this time, at least, it will be able to resist starvation. The locust's body water, dependent as it is on the state of the vegetation, is so exceptionally variable that its content can drop by approximately half yet still enable it to search for a new source of supply.

Also inside the body are a number of gossamer-fine silvery tubes. These are the trachae, or air tubes, and if examined under the microscope would be found to connect with a row of small apertures in the sides of the thorax and abdomen enabling the creature to breathe. The three main parts of the thorax are supported on three pairs of legs. Finally, if the locust is a female she will betray the fact by a pair of short black curved hooks protruding from the end of her 'tail'. They are her ovipositors and she will use them for digging in the soil before egg-laying.

In terms of muscular activity in proportion to body size, a flying locust works ten to twenty times as hard as a human being running at top speed. This fact alone indicates its superb powers of flight. Usually we think of them as being exercised in a swarm; but it is not the swarm only which must be able to move long distances from one feeding ground to another. For the solitary Desert Locust the food supply of the wadi also dies out, and then it too must shift. Because the solitarious phase is generally only uppermost when it is far removed from the ordinary haunts of man, many of the locust's movements in this condition are still unmapped, but it is pretty certain that the migrations of these lonely creatures are nearly as great as those of their fellows in swarms. That they prefer to fly by night, when predators are fewer, is also certain, for we found many on the move when our camp fires were lit.

Many more adventures must be made into the remotest corners of the desert, and much more learned, before the full story of the Desert Locust's life cycle in this key phase can with assurance be written down in detail. The men engaged in the surveys are like military scouts who must report every observable movement and behavioural trait of an often invisible enemy, so that little by little the elements of the picture in the field can be put together with those discovered in the laboratory. Nor can the commanding generals of the operation, the experts and scientists, afford to remain chairborne. It is doubtful if there is any other family of insects in the world whose studies have compelled so many dedicated men and women to travel so far, so long and so uncomfortably as the whole family of locusts. Much of their work, moreover, has of necessity had to be done in plague periods; so it is to the gregarious phase, in its most acute form, that we must now turn for a picture of the Desert Locust's life cycle in general.

Often, since I have returned from the desert, I have been asked how long an individual locust lives. The answer depends on where and when you find it. In one winter generation whose birth, migration, breeding and death were tracked from wadi to wadi across the deserts of eastern Ethiopia, the incubation period lasted 14 days, the hopper period (from the hatching of the eggs to the beginning of adulthood) 38 days, the immature stage 45 days and the breeding stage, ending in death, 30 days (all figures approximate). So there was a total of 127 days, of which more than two-thirds were passed in a state of sexual immaturity. This is the key factor in the locusts' longevity. The quicker they mature the shorter their lives.

The duration of this betwixt-and-between state depends on whether the fledgling locust meets with a run of warm-to-hot weather or warm-to-cold, for being a cold-blooded animal it is highly sensitive to temperature changes, which can affect its behaviour in many ways, including flight. The lower they are within a given range of tolerance, say from 125 degrees to a little under freezing, the slower the development. Under laboratory conditions a Desert Locust would normally be ready to copulate about three weeks after fledging. In the wild, however, it may do so within as little as a fortnight or as much as six months – a variation which can work out very much to its advantage at plague times, when its invasions extend from the burning wastes

3 Desert Locusts copulating. After fertilizing the female, the male often remains on her back while she searches for a laying site and lays.

4 *a* and *b* After using her ovipositor to test the site for moisture, the female thrusts it deep into the sand, where she deposits a pod of about one hundred eggs. Above this she excretes a foamy substance through which hatchlings scramble to the surface.

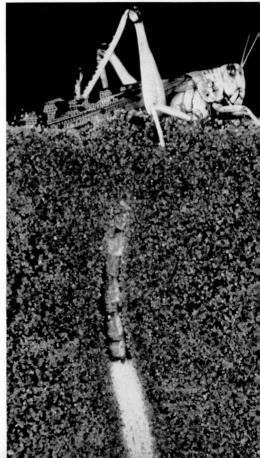

of the inland deserts to mountains and coasts which may be comparatively cool or even cold. (The desert itself can, of course,
present both extremes.)

Thanks to this elastic development, so conveniently geared to
the kind of temperatures the locust is likely to meet, a swarm born
in the wadis draped like dusty necklaces south of the fretwork
crags of the Tassili n'Ajjer in south-east Algeria, for example,
could migrate for 1,500 miles or more through the summer
Saharan heats to the lee of the Moroccan Atlas mountains and,
having overwintered in some cool valley, still push on afresh and
immature to begin the dreaded spring breeding which would
threaten all of the fruitful North African Mediterranean littoral.
(The winter pause also gives man a chance to destroy the invader;
unfortunately, in the past, he has not always been alert enough to
take it.)

Such an immature swarm is recognizable through its pinkish
colour; but with sexual maturation this begins to change to yellow.
Very often rain seems to play a part in triggering a mass impulse
toward maturity which rapidly spreads through the swarm. This
is something which for a long while puzzled scientists. For months
perhaps, in a very arid region, the insect may have lingered in a
state that in a human being we should regard as protracted pre-
pubescence, its colours unaltered, the female's ovaries undeveloped,
while both sexes, mutually indifferent to each other's charms, are
content to browse together in the withered vegetation. Then – not
with the rains, but just before them – comes the tell-tale colour
change which has been attested too many times by too many
observers for there to be any doubt about it. The pink vanishes
first from one part, then from the other, to be replaced by fawn or
yellow; and by the time the pink is almost gone the insect is ready
to copulate. In the field this gradual but quite quick transformation has been witnessed again and again, and when the rains have
come breeding has immediately followed. Nor has this happened
in one district only. In the Somali Peninsula, where the delay
between fledging and egg-laying is quite commonly a matter of
three to five months, these sudden outbursts of sexual maturation
and egg-laying have been known to take place more or less
simultaneously at sites separated by as much as 450 miles and

D

many other places in between; and invariably this has coincided with the advance of the rains. It is almost as if the locusts knew that the rains were near; yet if they could do so, this would not of course explain the simultaneous ripening of their gonads. Something then must trigger it off; but what? Humidity, temperature and the state and length of the available sunlight have all been suggested and found wanting as explanations. There is also a negative factor, in that a continuous diet of sere and yellow plants, as fed to locusts in a laboratory, can hold back maturity by up to nine months – less perhaps in the field, but the effect would be the same. Here again, however, we are up against the difficulty that the delay ends and copulation begins before the rains can possibly affect the issue. Except for a few succulents and aloes – which locusts do not eat – the perennial ground plants are still dry at this stage, and the annuals have yet to appear.

Among the more curious of the facts that have emerged from laboratory investigations into the whole field of sexual maturation is that it can be sparked off earlier if a locust has suffered from a traumatic experience of some kind. If, say, it loses a leg at fledging, then it will be compensated by achieving reproductive capacity sooner; electric shocks and being tumbled around in a revolving test-tube have also proved remarkably stimulating, but since neither of these is liable to occur on a widespread scale in a desert, other reasons have had to be looked for.

In the uplands of the Somali Peninsula it has been noticed that just before the autumnal rains arrive and when the skies are already beginning to be overcast, the air suddenly becomes fragrant with the scents of certain shrubs. All over the arid landscape, where most of the ground vegetation has yet to glimmer green, these shrubs, cued by a drop in soil temperature, begin bursting into leaf. Within about a week at the beginning of October the baked hills become a backcloth for a spatter of tiny blossoms and new leaf-clusters to a height of twenty feet. These are of little attraction as food and certainly play only a marginal part in the locusts' diet. On a particular occasion described by Dr Peggy Ellis of the Anti-Locust Research Centre, who spent three consecutive autumns in the peninsula observing swarms and collecting plants for Kew Herbarium, the bud-burst of the shrubs began in the first week of the month, the rains broke on the twelfth day, copulation began on the fourteenth, followed by egg-laying eight days later, and not until three weeks after the bud-burst was the ground cover green.

This sequence was common throughout the area investigated – about a thousand square miles. Wherever the observers looked, egg-laying seemed to have been more or less simultaneous. During the whole period when the parent locusts were maturing they were feeding on the same dried-up vegetation, and since maturation is not an instantaneous process – it takes at least seventeen days for the locust's ovaries to develop from their dry-season state – this left only the bud-burst as a possible pointer to a common factor by which the process was triggered off.

In the laboratories of the Anti-Locust Research Centre a series of experiments was prepared. Pieces of the resin exuded by some shrubs, and 'sticky-buds' of others were dropped into cages containing ten or a dozen young adult locusts. Although they remained uneaten, their aroma clung to the cages; and the locusts began to change colour. After essential oil from one of the shrubs had been spread on insects kept in solitary confinement the same thing happened. By the twenty-second day all of them so treated were significantly more advanced in colour change than the controls. The females in four of the experiments were then sacrificed and dissected and their oocytes (developing eggs) measured. Those of the treated animals were significantly longer than those of the controls.

The most effective of the essential oils used was obtained from species of a shrub called *Commiphora*, common in eastern Africa, and the constituents giving off its scent are known as terpenoids. In a final series of tests, therefore, pairs of locusts were smeared with a selection of the terpenoids and allowed to lay. The following result is recorded by Dr D. B. Carlisle and Dr Peggy Ellis, who carried out the experiments.

Egg-laying started in all groups except the controls between the 20th and 23rd day, and every individual female had laid a pod by day 24. In the controls the first egg-pod was laid between days 29 and 31 and not until day 47 had all the control females laid. By day 29 all the non-control females had laid a second pod; the second pod in the controls was laid between day 40 and 64. *In other words, every single treated female had laid two egg-pods before the controls had laid a first.*

In case it should be thought that the speeding up of the locust's maturity might be in some other way neutralized, they counted the number of egg pods in each case. Varying between 40 and 85, they showed no significance as between the treated locusts and the untreated insects used for comparison. Over ninety per cent of all the pods proved to be fertile.

There seems little doubt, therefore, that the locust shares with man a marked response to aphrodisiacs. The curious thing, moreover, is that they are the same aphrodisiacs well known in the days of the Eighth Plague and afterwards: myrrh (with which the ageing Emperor Tiberius rubbed his beard) from the shrub *Commiphora myrrhae*, frankincense from *Boswellia carteri*, and Balm of Gilead, the 'sticky-buds' of another *Commiphora*. All these grow in Eastern Africa.

However, it is a far cry from a laboratory in Kensington to the lonely corners of the deserts where *Schistocerca* is apt to breed and it is unlikely that the appealing scents of bud-burst explain the whole or even the major part of the maturation mystery. What these fascinating experiments of Dr Carlisle and Dr Ellis have demonstrated is that under controlled conditions the Desert Locust, when introduced to certain aromas, matures more quickly. In the wild what probably happens is that with the onset of the rains the whole environment begins to change and the *combined* effect triggers off the flow of the insect's hormones.

That nature thus works with the locust over a vast area is shown by a large number of examples tabulated to cover the whole period of the last plague. They show that synchronous egg-laying has occurred in countries in almost every breeding area between India and the Atlantic and in almost every month of the year, depending on the onset of the local rainy season.

Just as the bursting buds of the aromatic desert shrubs foretell the rainy season and work mysteriously on the gonads of the locust, so does the rain itself fulfil the ensuing need by providing the soil conditions in which eggs can be successfully laid and in-cubated. We are here, in fact, at the heart of the whole life cycle in which every stage is marvellously interlocked with every other in such a way as to reap advantage from the briefly favourable aspects of an otherwise hostile environment.

Copulation begins almost simultaneously with the beginning of the rains and is accompanied by a colour change. The male, the more brilliant, is ready to copulate when yellow shows in the tip of his abdomen. He is mature, as a rule, before the female, who per-sistently refuses him, throwing him off her back until the dis-appearance of pink from her hind tibia shows that she too is ready. Her own maturity, meantime, is being hastened by a skin odour given off by the male. The odour also affects other males, so that the process of maturation of the whole swarm, numbering many millions, is speeded up and will usually be complete within a week. It is believed that vision, the most powerful factor in holding the swarm together, also plays a part in this sudden burst of sexual unanimity.

Having at last achieved a successful leap on to a female, the male holds her with his front legs while performing an act of copulation which may on occasion take from three to fourteen hours but on others may be repeated on different females several times in an evening. Often, the act completed, he remains on her back while his mate, becoming restless, begins to move about. This is not necessarily due so much to annoyance with her incumbrance as to the need to find a suitable egg-laying site. It is now that the hard hook-ends of her ovipositor come into use, digging and probing into the sand for several inches as she tests it for humidity. A water-logged site will be rejected, as will a very dry one, though if necessary she will bore through a couple of inches of quite dry sand, using the delicate sensory apparatus in her ovipositor, before making her decision. She can also use it to test for salinity. Although the seashore is no use to her, she may lay a little way inland, above the line of salt-impregnation.

The search as a rule goes on for one and a half to two hours and may involve several rejections. Other females join her, possibly moved by the assumption that if one finds a good site the rest are bound to do so also, though laboratory tests suggest that smell is the mutual attraction rather than any primitive form of instinctive reasoning. (In cages a tethered female was soon surrounded by other females; so were rolls of paper that had been in the cage and picked up the locust smell.) The value of such gathering together is that it is yet another factor in maintaining the swarming locusts' cohesion, for naturally if the females are clumped so are their egg pods and this patchy distribution means that the resulting young locusts, in their hopper form, start life in a number of groups which can rapidly band together.

The acts of probing and egg-laying are fascinating to watch, if only for the locust's extraordinary ability to extend her abdomen, which seems as if it must be made of elastic. Down it bores and down, searching for the desired conditions at depths up to three times the normal length of her body. Bending it gymnastically she drives the hole at a right angle to a depth of from four to six inches, where the temperature of the soil, even in the most baked of deserts, is noticeably lower than at the surface. This is important in her battle to preserve the eggs from desiccation – they would otherwise be dried out and killed. Some may in any case suffer this fate if within three days of the eggs' full development in her body she has been unable to find a suitable laying site, for she then has no choice but to eject them, either on to the soil or the branches of bush or tree. This is one of the causes of plagues dying out. It can happen when there has been just sufficient rain to encourage breeding but not enough to moisten the ground sufficiently for it to be carried through successfully. A whole swarm may be thus affected. But if the females have found the conditions they want they are likely to leave behind them a patchwork of closely clustered pods whose egg total may run into hundreds of millions. And then the plague will not merely be maintained but intensified.

Scientifically considered, a locust egg pod is a rather beautiful thing. Having probed and dug, the prospective mother retracts her abdomen slightly and exudes a frothy substance which is partly absorbed into the soil of the hole, hardening it to prevent any particles falling in. At the same time she begins to pump the eggs through the opening in her ovipositor and as these slip out they rotate in the bottom of the hole in such a fashion that when finally at rest the head of the future embryo points upwards, the way it must go when hatching. More froth is emitted with each egg-laying, so that as the pod builds up, the walls are systematically secured. Finally, when about eighty eggs have been laid (by locusts in the gregarious phase, more when they are solitarious), the rest of the hole is filled with foam which hardens just enough to keep it open. Through this softly brittle core the infant locust, in due course, emerges as through an escape hatch – assuming, that is, that he has in the meantime escaped certain perils. Parasites or fungoid disease may kill off a whole egg-field, or it may suffer through drying-out of the sub-surface moisture.

A thing surprising to the novice in the desert is that even in the most arid zones the sands of the wadis are seldom absolutely dry about four or five inches down, though a hygrometer may be needed to detect what moisture there is. The female locust's hygrometer is her ovipositor but, efficient as it is, it cannot assure her of every requirement. Ideally, the egg should be able to absorb from the soil its own weight of water within the first five days. The occasions when there is not enough are not, however, necessarily fatal. Cases have been reported of eggs lying dormant in dry soil for weeks, neither gaining nor losing much water until a fresh fall of rain has caused a spurt of development. If, in the meantime, there has been renewed laying in the same area, there is a danger that all the eggs will come on together, causing a piling up of the locust army and a reinforcement of the plague.

The sight of a large egg-field hatching out is one no observer is likely to forget. One who had this experience in East Africa, the entomologist Clifford Ashall, said: 'So far as the eye could see they were boiling out. Every inch of open ground appeared to be bubbling young locusts as pod after pod, many of them only a couple of inches apart, gave up its contents. This went on for three hours, from just before dawn. Within three days the whole of a vast egg-field had been hatched out and the hoppers were ready to march.' What sets off this terrifying bout of simultaneous hatching no one really knows, though the fact that all the members of an egg pod emerge so rapidly one after another has suggested to some locust workers that there is a mechanical stimulus at work – that one insect, having hatched, bumps into the others as it tries to wriggle up to the surface and so sets them all hatching. Whatever the cause, the result is another demonstration of the Desert Locust's blind drive towards the utmost cohesion when in the gregarious phase. By emerging in such numbers it has a fair chance of swamping its predators and clearing the way for the formation of the marching bands which prelude the flying adult swarm.

All over a newly hatched egg-field, when a light breeze has risen, tiny dry whitish balls are to be seen blowing over the bare patches between the herbage, catching on twigs and freeing themselves to bounce on again. They are the rolled-up skins of the hatchling's first moult, thrown off almost as soon as it has emerged. At this stage it is about half an inch long, a minuscule creature, mainly black, in which a pale colour pattern steadily becomes more marked as it nears the next moult, taking it into its second instar (the period between the moults). Five moults are necessary before achieving adulthood. The young locust hopper is in effect exchanging its suit of armour to accommodate its growing body. As the old one is discarded the new one below it expands and then hardens to protect it; so at each stage of growth the process must be repeated.

Between the first and second instars there is little change except to the eye of the expert, who notes the intensified coloration. With the third instar the wing buds can be seen, a small wing overlying a larger one. At the next stage they are inverted from a lateral to a dorsal position and become well marked and distinct. The position is reversed once again at the last moult. All this takes from three weeks to two months or more, depending on local temperatures.

As the last instar ends the hopper is ready to become a locust. Climbing a suitable twig, as in previous moults, it hangs itself upside-down and begins a pumping action of the body. The cuticle splits behind the head, the split opens forward and gradually the insect forces itself out while hanging on by its forelegs and still pumping. When almost out it bends up, catches on to the twig by its forelegs and levers itself out of the rest of the cuticle. Still hanging under the twig but now holding on by its forelegs, head upward, it dries itself and begins pumping blood into its wings, preparing them for the many hours daily it will in future spend in flight.

Chapter 4

The Phase Change

To speak of the insect as being in its 'solitarious' phase or, alternatively, its 'gregarious' one might seem to imply a hard-and-fast
division between two ways of living and general behaviour. But,
of course, nature is seldom so absolutely divisive. Extremes are, on
the whole, abhorrent to it. Even in the worst 'years of the locust'
there is likely to be some time or place at which the insect seems
as if poised uncertainly between two opposing pulls, one toward
solitary living and the other toward swarming. Looked at over
longer periods, however, the differences can be seen as very much
more distinct. They will be shown not only in behaviour but in
colour and even, although to a less certain extent, in the locust's
morphometrics – the measurements of its physical form. For
perhaps years it may remain one of the world's most withdrawn
creatures. Innocent and harmless-seeming, more victim than
menace, it lives an isolated life of great difficulty, usually in haunts
far from those of men. Then, for reasons to be seen, comes a
series of population upsurges leading to a break-out and in a
brief time, if unchecked, the locust has become the scourge we
know.

Such extremes of behaviour within one species must seem almost
inexplicable – only the human species, indeed, can rival it. Nor
does the nomenclature employed to describe this insect clarify the
picture, for how can a species named *Schistocerca gregaria* be also
solitaria? – and how, for that matter, can it be 'solitary' or even

43

'solitarious' when, while in this condition, it can be found in ten of thousands or even hundreds of thousands, all in the same stretch of wadi? One must sympathize with scientists in their hunt for the right word, for this ability of the locust to live two lives, each apparently contradictory of the other, is one of its most mystifying characteristics.

The first entomologist to realize that these extreme forms represent two aspects of the same insect was Dr Uvarov. The object of his studies was not, in the first place, the Desert Locust, but a relative, *Locusta migratoria*, the Migratory Locust, which in 1912 had heavily plagued the Northern Caucasus where Uvarov was working.

While working on the Migratory Locust he also gave his attention to another and apparently quite harmless locust, *Locusta danica*, then regarded by most entomologists as a creature of a separate species. Their error was understandable because *danica*, causing so little trouble to anyone, had not been much studied and little was known about its life cycle and ways, except that, unlike *migratoria*, it seemed to have no particular permanent preferences as to habitat. Almost anywhere in the vast grasslands of Europe and temperate Asia suited it equally well. *Migratoria*, on the other hand, liked to base itself among the jungle-like thickets of gigantic reeds, often ten to fifteen feet high, growing beside the ever-changing channels of the great river deltas of the Black, Caspian and Aral Seas and Lake Balkhash. For breeding it chose grassy isles and banks where the reeds had not yet grown.

In colour and behaviour the insects were also markedly different. For instance, if a wandering swarm of *migratoria* hoppers came across a lonely specimen of their kind it would promptly join the band, whereas, if an isolated *danica* found itself overtaken by a swarm of *migratoria* it would leap to get out of the way.

'I must candidly confess,' wrote Uvarov later, 'that when starting my work I had only a very modest intention – to find out characters for separating *migratoria* and *danica*, which I assumed beforehand to be distinct specifically. The facts quickly destroyed my preconceived opinion.' For it now began to appear that in spite of every superficial manifestation of unlikeness the two locusts were one. Field work showed, moreover, that under certain circumstances, not then clear, they could so modify their differences that the two forms were seemingly interchangeable.

What were these facts, as they were revealed later in the laboratory cages? It is important to understand them, for, with qualifications, they still underlie much of the present practice in the approach to the locust problem. Boiled down, and for all their superficial differences, the only trustworthy criteria by which the scientist could separate the retiring *danica* from the marauding *migratoria* lay in two sets of physical measurements. These involved the shape of the pronotum (the hard shield-like covering of the fore part of the thorax behind the neck) and the ratio of the length of the wingcase (elytron) to the hind femur. Many other considerations had to be made, but it was from this point that Uvarov began the deductions leading to a revolutionary theory.

In carrying out control operations, following a great invasion of locusts into the Stavropol province of Northern Caucasus in the autumn of 1912, Uvarov also remarked that, whereas the first swarms were undoubtedly *migratoria*, some of their progeny were equally certainly *danica* when judged by their coloration and a marked tendency to desert the swarms. This showed up clearly in their third stage as hoppers and still more so when they were adults. Even more numerous than these obvious changelings were specimens of an intermediate character which could not be identified with either *migratoria* or *danica*, except in their morphological measurements. These seemed to show that as the swarms grew fewer and more scattered as a result of control measures, the numbers of the *danica*-type grew greater until finally many had lost their gregarious habits altogether and were scattered all over the steppe in pure *danica* form. Unfortunately, he had not been able to secure reliable examples from more than two such generations when he moved from the Northern Caucasus to Tiflis to set up and direct a plant protection organization as one of a regional network he had advocated and planned. This was not concerned with the locust problem and indeed it was not until the end of the Second World War that he was to return to full-time locust work. But the germ of his revolutionary theory of the phase change in locusts had been born. As eventually amplified, it was to explain, for the first time, how the great plagues, not only of Migratory Locusts but of all locusts, occur. An unknown British army cook, quartered at Tiflis during the First World War, and an outstanding British entomologist, Dr Patrick Buxton, on leave there, helped forward its fruition.

'The British soldiers in Tiflis,' Sir Boris recalled, 'were on their own and feeling rather lost, so as my wife and I spoke a little English they used to come round to chat. Before the war I had been in touch with the Imperial Bureau of Entomology on scientific matters and when the cook was going on leave I asked him to take a letter applying to be able to work there, which he very kindly did. Buxton came to Tiflis on short leave from Mesopotamia, chiefly because of its very good museum in which I was then working. There we got acquainted, with the result that he wrote a letter recommending me and supporting my application.'

Meantime Uvarov had been struck by some resemblances between the 'Russian' Migratory Locust and its formidable counterpart, the African Migratory Locust (*Locusta migratoria migratorioides*), whose outbreaks had periodically laid waste to vast areas in West, East, Central and Southern Africa. In London, where he arrived in 1920 to join the staff of the Imperial Bureau of Entomology, he began to collect evidence suggesting that the African insect was also subject to a phase change and that here again what was being reported from many countries as two species of locust was, in truth, one. The importance of the discovery was that it pointed the way to a new approach to control strategy, for if, in its author's words, 'the wonderful phenomenon of the transformation of a swarming locust into a solitary, harmless grasshopper' explained the periodicity of locust invasions – and recessions – then evidently the best way to check invasion was to stop the enemy in its non-swarming stage, that is, in its outbreak area. This, however, first had to be found.

If this seems self-evident now, it was not so at a time when great areas of Africa, and those most affected, were at best sketchily mapped, presented prodigious difficulties in travel and enjoyed the plant protection services of only a handful of entomologists, all of them drawn from Europe. Meanwhile, in the far south of

Africa, another distinguished entomologist, J. C. Faure, was coming to much the same conclusions as Uvarov, but about a different species, *Locustana pardalina*, the so-called Brown Locust, which had been thought to originate in the Kalahari Desert, only invading the Union when swarming. He wrote to Uvarov:

My personal experience with the species began in the summer of 1914-15, when scattered swarms began to appear shortly after the break-up of a prolonged and very severe drought. Voetgangers (i.e. *nymphs*) of all stages and flyers occurred together in loose swarms, and it was practically impossible to destroy them by the usual method of poisoning. The swarms did not move in the usual compact formation, nor did they camp for the night in dense clusters. Many of the adults were strikingly undersized (a characteristic Uvarov had noticed in *danica* as compared with *migratoria*) and a large percentage of both adults and voetgangers were abnormally coloured. Only in swarms that approached the normal in density did the typical orange and black colours of the voetgangers begin to show up.

Although I did not realise the fact at the time, I was witnessing the transition from the grasshopper to the swarm phase. Toward the winter, that is in May and June 1915, the flyers began to move about in fairly definite loose swarms, and they laid their eggs in compact deposits, with the result that large swarms of typical swarm voetgangers hatched the following spring.

This was at a time when no reports had been received of invading swarms coming in from the Kalahari or anywhere else. But scattered locusts *had* been seen in the infested area in the previous summer and Faure was forced to conclude that, properly speaking, there had been no invasion at all but that the swarms of September-December 1915 had arisen from locusts bred up inside the Union's own borders, without outside help. In 1917 another severe outbreak of voetgangers followed the appearance of scattered locusts earlier in the year, and this time Faure, realizing what was going on, spent a good deal of his time on the veld, capturing and closely examining both single and swarm specimens.

He now became sure that *pardalina* not merely existed in

scattered swarms or compact swarms – it also lived as a grass-hopper, often miles away from the nearest swarm, and (as he later assured himself) in seasons when there were no swarms at all. 'The specimens captured singly,' he noted, 'almost always have the colours of the grasshopper phase, and they are as a rule a good deal smaller than swarm forms.' This could occur either as nymphs or adults. Their colours, moreover, varied widely from the universal brown, when swarming, which gives *pardalina* its name.

It would be quite an undertaking [he continued in his letter to Uvarov] to describe in detail all the shades of colour they exhibit. Usually there is a striking protective resemblance. Where there is plenty of green grass both voetgangers and flyers may be almost entirely green, or at least partly green. Where the veld is only sparsely covered with grass and bushes, they resemble the colour of the soil more or less. In parts of the Karoo, notably Beaufort West and Prince Albert, there are patches of gravelly soil varying from slatey-blue to almost black. In 1917 I was very surprised to find a very striking ten-dency among the scattered voetgangers to vary in colour from place to place more or less in accordance with the colour of the soil. When the progeny of these locusts appeared on the same farms in swarms six months later there was no trace of such a protective resemblance: they all wore the King's regulation swarm uniform!

With these completely independent observations by Faure to support his own, Uvarov was more than ever confident that his phase theory need not be confined to one or two species but would be found occurring in other members of the locust family. Put briefly, his conclusion was that the lives of all locusts can be divided into phases in which the differences of form may be slight but are nevertheless measurable, and that these differences may, in general, go along with differences in behaviour. The differences in both cases will vary from species to species, being small in some and great in others, and there may be many stages between the phases, when the insect seems to be in transition between two distinct personalities.

This is basically the theory which, with modifications, has now stood up for half a century, since Uvarov expounded it in the September 1921 issue of the *Bulletin of Entomological Research*. However, eight years had to go by before it was established in the Desert Locust.

Here again most entomologists had thought that there existed not one but two species, one being listed as *Schistocerca gregaria*, the other as *Schistocerca flaviventris*. All the obvious characteristics of *flaviventris* seemed to suggest this: its remote desert habitats, the way one insect would shun another, its less active behaviour, its protective coloration – they all added up to a very marked difference from the locust of legend. On the other hand, they also squared with Uvarov's contention about locusts generally, that for reasons not clearly understood but amply demonstrated already in *migratoria* and *pardalina*, there was a biological swing between extremes. The difficulty in proving it in *Schistocerca*'s case was that, as *flaviventris*, or in what would now be called its solitarious phase, it tended to disappear into the unknown almost completely. There could be no tramping about from farm to farm, watching it in the process of change. What was wanted was a camel and a great deal of time.

H. B. Johnston, the government entomologist in the Sudan, had one but not the other when he made a journey up the Red Sea coast north of Port Sudan in the intense heat of July, 1929.

I was not at all interested in locusts [he now recalls]. My first sight of them *en masse* had been during the War, when I had been drafted as a soldier to the Sudan and saw an immense continuous and very smelly heap of them lying on the Red Sea coast, where they had presumably foundered during flight. In 1929, I was looking for areas along the coast with enough rain or flood to be suitable for raincrop cotton – what was called American short staple – to supplement the fine Egyptian irrigated cotton being grown at Gezira. The coast is generally very dry but there are wadis coming down from behind and opening out into wide deltas. If there is sufficient rainfall the ground retains enough moisture to bring up the crop, and with cotton booming it was thought this could be profitable. My job was to study the deltas for possible cotton pests, not locusts. The going was too hard for a car – nothing but salt bush for mile after mile and hour after hour. Every so often as the camel put his foot down out would jump a locust. They were green. I decided to collect some and when I got back to Port Sudan I sent them up to Uvarov in London, where he labelled them as *Schistocerca flaviventris*.

After the rains had begun I returned up the coast, again on the cotton. This time the whole place had become a savannah, and there I found *Schistocerca flaviventris* breeding, but not producing green hoppers as it normally would but another hopper altogether, one which became bright yellow. This surprised me very much. I was so interested that I am afraid that for the time being I forgot cotton and concentrated on locusts. There they were, showing all the signs of the phase change. I had Uvarov's theory behind me and now I was seeing it in a species to which it had never before been applied. Here we were having what he said might happen. First, the solitary locusts I had seen on my previous trip; then, through one or two generations, the production of regular gregarious hoppers which had bred from them. There was the transition right in front of us. So *Schistocerca flaviventris*, which we had thought of as a separate species, was now found to be the solitarious phase of the Desert Locust. One couldn't do very much experimentation out there, however, for they were already beginning to damage the cotton and I had to get busy controlling them. However, I took some of these gregarious hoppers and put them singly into cages. This somewhat surprised an official from Port Sudan who went back to Khartoum and said, 'You know, that fellow's not killing locusts at all – he's breeding 'em!' What we were able to do in fact was watch the effects among survivors when large masses of the free locusts were killed and compare them with those in the cages. We were able to observe that if they were in their last instar when caged they produced a gregarious locust of the expected colour, but that, if they were very young they gradually, while moulting, turned green again. The survivors in the cotton, after the drastic thinning out caused by the sodium arsenate poison bait, also, if they were young enough and not in the final moult, began turning back to the solitary colour. It

5 Hopper becomes locust (1). Climbing a suitable twig in preparation for its final moult, it hangs itself upside-down and by means of a pumping action splits the cuticle imprisoning it. When almost out, it bends upwards, catches the twig by its forelegs and levers itself free.

6 Hopper becomes locust (2). Head again uppermost, the fledgling dries its wings and begins pumping blood into them, preparatory to short flights.

was a marvellous panorama of the whole phase theory at work over a period lasting from July of one year to March of the next. This was when it was realised that the first step to controlling the Desert Locusts must be to watch the solitary hoppers and if possible take action before they became gregarious.

Johnston's eye-witness account of the miracle of the phase change in yet a third species, gratifying as it was to Uvarov, raised, as he realized, many more questions than it answered. What were the fundamental causes of this mysterious shift from one extreme of the locust's living habits to the other? Why at one time so solitary and shy and at the other so gregarious that the earth and every plant and tree in it seemed to pullulate with locusts? What triggered off these fantastic population explosions and, alternatively, caused their recessions? What were the stimuli of the changes in the individual locust? What mechanisms, once the swarms were in being, governed their cohesion so that for thousands of miles they appeared to be able to stay together, sweeping over continents in clouds as vast as those of any storm? And how, important above all from the point of control, could their movements be forecast?

In his 1921 publication Uvarov noted:

The direct cause of [this] ignorance is that injurious insects, and locusts especially, are studied only in the years of maximum development, and nobody cares about them in the minimum years, when the clue to the whole locust problem is likely to be found.

Had his warning been generally heeded it is unlikely that the subsequent plagues, lasting for thirty-two years out of the fifty since his theory was published, would have been prevented, but they could conceivably have been checked in a degree that would have saved many hundreds of thousands of poor peasants much misery. Thanks to Uvarov and the band of dedicated men and women he led and inspired, and to a number of French, Indian and Egyptian scientists of the same ilk, some at least of the answers are now known.

Many of them are contained in Sir Boris's own opus, *Grasshoppers and Locusts*, Volume I, a definitive work whose publication in 1966

E

coincided with his diamond jubilee as a working entomologist. Eleven years after his retirement from the Directorship of the Anti-Locust Research Centre, this remarkable and infinitely industrious man was still working on his second volume, covering the whole range of field studies and control, when he died in March 1970 at the age of eighty-one.

Chapter 5

Saharan Panorama

In January 1966 populations of Desert Locusts everywhere were apparently at a very low ebb. After surging to and fro for thirteen years over practically all the territory between the Bay of Bengal and the coasts of West and North-West Africa and as far south as Tanzania, the swarms had to all intents and purposes died out in 1962. The decline was gradual, first in the west, then in the centre and finally in the east of the invasion area, where a heavy aerial spraying campaign by the Pakistanis and the Indians probably gave them the *coup de grâce*. In the past such lulls have been mainly occasions for thankfulness enabling farmers to pursue their hard toil without added worry, while government plant protection officials turn their attention elsewhere. Yet it is precisely at such times that study of the locust and its ways is most essential. During plagues all hands are engaged in fighting it. Research continues in the laboratories and where possible in the field but necessarily most of the effort is concentrated on the tracking of swarms and hopper bands and improvements in methods of controlling them. The mysteries of the phase change, however, are best probed between plagues, when the Desert Locust, diminishing its threat, has retreated into the marginal lands and desert wadis where it will remain until circumstances again encourage it. However small its numbers at these times and however remote its hiding places, it must then be sought out and studied, for only in this way is it possible to construct a meaningful

53

picture of the nature and development of the behavioural traits which could lead to a new outbreak.

Between 1958 and 1963 immensely valuable work in this respect had been done by George Popov. His lifetime's experience in Desert Locust control campaigns, and his prodigious journeys, produced many important consequences, one being that he established for the first time, in great detail, the kind of habitats favourable to Desert Locust breeding and where, *in general*, they were to be found. Unhappily, his FAO/UNDP Ecological Survey also confirmed beyond a doubt that the insect has no clearly defined breeding areas like that of its cousin, the African Migratory Locust, which is faithful to the flood plains of the Middle Niger in the neighbourhood of Timbuktu and can therefore be controlled there. The Desert Locust, Popov showed, can break out wherever and whenever conditions are suitable, and these may occur in many places. Notable flare-ups in the past have started around the Persian Gulf, the Indian Ocean, the Red Sea (both coasts) and the Sahara. To search all five million square miles of such country would obviously be impossible, but Popov pointed out that although geography is of no help in pin-pointing the Desert Locust during recessions, topography can play a significant role, for it is on the bare rocky uplands of the deserts that rain occurs most frequently and around these, therefore, that vegetation of a kind fulfilling the locusts' needs is most likely to be found. He listed a number of areas with major topographical relief features shedding water, at infrequent seasons, far into the arid lands below them. One of these was the Hoggar Mountains, whose sombre peaks and precipices dominate the central and southern Sahara.

At the end of 1965 I had been invited to join in three long-distance searches financed by the United Nations Development Programme, to be made in Algeria, Saudi Arabia and Iran. They were to some extent pilot projects for the system of national and international surveying which now operates throughout the breeding zone. The Algerian trip, crossing the Sahara by Land-rover from the Hoggars to the borders of Mali and Niger on one side and of Libya on the other, turned out to be the most rewarding in terms of locusts discovered, although its significance at the time was not fully apparent either to me, or, I think, my companions.

I arrived in Algiers at the end of the first week in the new year and was met by Professor Roger Pasquier, a veteran zoologist and member of the Algerian Agricultural Institute who advises the government on locust problems. Tall, slightly bent, with an initial reserve quickly cracking open when he finds a receptive listener – but impatient of fools – he looked with some gloom at my wind-cheater and pronounced it insufficient for Saharan nights. He was himself wrapped in an incredibly ancient overcoat and wore a beret tugged down to his ears. A bluff young Arab, Hadj Benzaza, a former pupil of Pasquier's, and now head of the government plant protection service, peered from a *djelleba*, a cloak-like hooded garment ideal for rolling up in when sleeping out at nights. We boarded a Dakota serving the oases. Lumbering up from the ground at dawn like an arthritic pterodactyl, she headed first out to sea to gain height before turning for the long climb over the peaks of the Tell Atlas Mountains hemming in the city and fruitful littoral on the south. The plane is a kind of air-bus, making a twice-a-week round flight across the Sahara. One flight (ours) goes more or less directly southward, calling at Laghouat, Ghardaia, Ouargla, El Golea and In Salah to Tamanrasset, a flying distance of 1,100 miles. This is the western route. The eastern route, on the other hand, makes a dog-leg after Ouargla and arrives at Taman-rasset via Fort Flatters, In Amenas and Djanet, taking a couple of days over it. Our plan, drawn up in military fashion by the Professor, was to disembark at Tamanrasset and return by the eastern route eight days later from Djanet. We would thus be able to make a sweep through the most important of the southern and eastern wadis where breeding could have occurred during the preceding months.

Our fellow passengers were a mixed bag, typical of rumbling old planes to be found in all the remoter quarters of the world: engineers, oilmen, assorted merchants and the inevitable party of tourists. What did we expect to see? Probably each cherished a different notion, for we knew from books of exploration that the Sahara is many deserts in one and that one might wander for 1,500 miles southward, or east and west for 3,000 miles, seeing little but the bare bones of the earth's surface, denuded rock, wind-heaped sand, parched mountains and great gravel plains where nothing moves except a rare camel caravan and, at intervals sometimes of years, the Desert Locusts riding the restless winds. This is their country, and the map showed its scale: a barrier stretching from the Atlantic to the Red Sea. But it is really larger than this, for the Red Sea, like the Nile, is merely an interruption. Beyond it the deserts roll on through Arabia and again, beyond the Persian Gulf, through Iran to Afghanistan, Pakistan and India. On the north and south sides geographers have had to be content with a meteorological contour which can only be approximately identified on the ground by changes in vegetation. This line is where the rainfall averages one hundred millimetres a year (four inches: England's average rainfall is 33·83 inches). And then there are the steppes, huge areas of stony landscape which in other parts of the world could well be called desert, for nothing of consequence grows there. The true desert, Pasquier explained, begins at the inside edge of the steppe, which may vary from a few miles in width to a hundred or more. Southward, merging with the steppe, is the savannah.

As yet all this was only an anticipation. It was not yet light when we cleared the first range of mountains. Already the plane had acquired the free-and-easy atmosphere of a flying picnic. While some of the tourists munched sandwiches others clambered over the luggage stacked forward, hopeful of photographing the sunrise. We flew at a height of about 7,000 feet, ideal for a panoramic view showing ground detail. Pasquier, in the seat behind me, offered an occasional commentary. He was sixty-five at the time of this journey and had travelled in the Sahara, often by camel, for more than forty years. He sketched a map in my notebook. It showed the Atlas Mountains stretching in a bow some 600 miles long from Morocco's Atlantic coast through Northern Algeria to Tunisia. They are significant in the Desert Locust story because their winter weather, often rainy and cold and sometimes snowy, forms a temporary temperature barrier to summer and autumn generations migrating from the south. In control operations this has an advantage because it enables swarms to be located and sprayed; but in the past it has led to enormous losses when the locusts settled in fertile valleys or oases where mountains blocked the way out. The Moroccan fruit-growing disaster in the winter of

1954-5 was one such case. With the onset of spring the locust can again move on and if allowed to cross the mountains can wreak its damage anywhere along the highly cultivated and densely populated North African littoral. Pasquier explained: 'It is swarms coming north through Morocco that most concern Algeria, for if they arrive when we don't expect them, or if our control measures are inadequate, they are likely to breed again. We could then get a northern spring generation, another big build-up of numbers, and the completion of a vicious circle, because, while some of this generation will continue eastward, others will eventually return south across the Sahara to the wadis from which their parents started. There men may in turn breed another tropical summer generation and the same pattern of movements will continue.'

Daylight had expanded over Algeria while he gave his little lecture. Directly below us the Tell Atlas heights gave a hint of their beauty in the form of dots and clumps of juniper and small evergreen oak. These gave way to swelling shapes of foothill downland, in parts much eroded, so that great gaps and rents appeared where there had once been forest. Small farms green with corn laid their pattern along the slopes and once I saw the snowy gleam of a blossoming almond orchard. For Hadj Benzaza, as well as for the locusts, this was frontier land, for it was here he would have to fight them if they succeeded in breaking out of their breeding spots in swarms. Organized into teams as directed by the local communal syndicate, men would be allocated vehicles, spraying gear and pesticides – chiefly BHC dust – at present held at seven sub-depots strategically located along the mountains between Oran and Bône. 'The nearest one to us now,' he said, 'is at Boghari, down there on the right. In store at Algiers we have 2,500 tons of powder and 250,000 litres of liquid chemicals. There are two hundred small powder-blowers for hand-carrying and sixty-five big ones which will be mounted on trucks. We shall have to hire a Piper for the air-spraying. Our job at Algiers is to supply extra equipment and extra men as needed and provide the over-all direction. If an invasion begins we shall send ten teams forward to support the syndicates. They divide the terrain into sectors. It's just like a military operation. First we shall lay down a barrage under the Saharan Atlas or wherever else swarms have been located along the front, then, if necessary, we shall fall back mopping up any that have got through. At least, that is how we expect to work. Fortunately we have not had to put the plan into operation.'

The rolling hills of the Saharan Atlas, parallel to the Tell Atlas, ended in a steep ridge overlooking Laghouat and we came in to our first landing on a desert airfield. But it was not yet the true desert, for Laghouat, although fairly claiming to be the gateway to the Sahara, stands among flat, sombre, stony steppe, a dreary sight from the air, interesting only for its *daia*. These extend over a wide region, probably a hundred miles or more across, stretching south of the town. Looked at from the air, they showed up as a number of rough circles, anything from fifty or sixty yards to several hundred yards across, some glowing green against the drab background of the steppe, some neatly ploughed and awaiting crops, some with little orchards, some abandoned to scrub. They reminded me of similar cultivated patches among the dry uplands of the French *causses* above the River Tarn. The Algerian *daia* are geological curiosities of the same kind, pockets of good soil holding enough moisture to permit a highly local agriculture where none would otherwise be possible. The farther south we flew the poorer they seemed until eventually one could only tell their existence by a huddle of stunted trees. All the *daia*, like the oases, are highly vulnerable to locust infestation. Dining at Ghardaia on our return journey, our host said that he had seen the insects three feet deep in the town streets, while the branches of date palms cracked under their weight with a sound like pistol shots. 'It was a horrible sight, as though a river were pullulating with locusts. They heaved, struggled and scrambled perpetually over each other and although we blockaded doors and windows many got in. There were stories of children being eaten by them, but I think they must have been babies who were smothered.' Pasquier said that many oases had similar stories to tell: 'When they pass on the swarms leave ruin, desolation and despair behind them. Livelihoods vanish and may never be restored.'

In the past much essential information about the movements of Desert Locusts has had to be gained from nomads whose long journeys and need to be perpetually on their guard against enemies, both human and natural, made them ideal observers of the most minute aspects of desert life. They would be the first to note the signs of egg-laying or to report the appearance of hopper bands. In the north-central Sahara most of them belong to the Chaamba tribes, whose encampments studded the steppe on either side of the road below us between Laghouat and Ghardaia. They were at this time just beginning their yearly round, moving out of their bases among 'The Five Towns' of the M'Zab. Here and there we could see the white patches of their flocks of sheep. There would also be goats and donkeys and all would be on the move from late December until midsummer, when they would set up temporary quarters in *zeribas* – grass huts – finally returning to their bases in time for the date harvest in the autumn.

Even glimpsed from a plane window, the walled towns of the M'Zab, each clinging to a hill, look incredibly romantic. Excepting Ouargla, the capital of the eastern oases, about a hundred miles south-eastward, they are the last big human settlements before the true desert opens. Ghardaia, largest of the five, gives its name to the airport. As we clambered into the sky again we looked down on grim bare rocky uplands. Beyond these, eastward, westward and southward, rolled two vast sand seas – the Great Eastern and Great Western Ergs, their dunes rippling into faraway obscurity under a thin yellow haze consisting of their own particles floating on the wind. In its way this airborne 'desert' seemed even more eloquent of the magnitude of the Saharan waste than the ground over which we were flying. Together, these two ergs – the name is Arabic for a collection of dunes – cover approximately 76,000 square miles of what once must have been tolerably fertile earth. The dust blown from them discolours the air of an area at least five times as great and may indeed drift for a thousand miles. (In West Africa the sun may be put out of countenance for weeks on end when the harmattan wind blows from the south Sahara.)

Only the Eastern Erg is traversed by a road made up along the line of an old caravan trail passing through a 200-mile-long gap, the Gassi Tawil. 'Gassi' means street and on either side of it long fingers of dunes stretch out from the main masses as though to clutch the vehicles serving the Fort Flatters oilfield. The Western Erg, 300 miles long, 100 to 150 miles wide, is immune to passage except by a dust track curling round the south of it. The tarmac road from Algiers ends in the lee of the erg, less than a third of the way across the desert. From the centre of the plane's moving shadow to the skyline the central Sahara revealed itself to be a giant jigsaw composed of dunes, flat yellowish soil – totally infertile and partly encrusted with salt – immense bare plateaux slashed by high cliffs and – standing up from the yellow – islands of dark grey rock eroded into strange whorls and fantastically twisted ravines as though a maelstrom of a prehistoric sea had swirled here when the rocks were young.

We had landed briefly at El Golea below the eastern edge of the Great Western Erg. The 250-mile stage to the next oasis, In Salah, found most of us asleep. Only Pasquier remained alert, watching through the window and passing notes and sketch-maps forward instructing me to note new curious rock formations and distant features. Once far off – I cannot now remember where – we saw the smudged smoke of a natural gas plant which anywhere else would have been an eyesore. Here it meant nothing, for the scale of the desert reduced it to a cigarette puff. Suddenly the plane was heeling over the snapped-off edge of the Tademaït plateau. We glimpsed a honeycomb of flat-topped houses, then a flurry of palms as we came in to land about three miles away. The heat of the sands met us with a slap as we stepped down. Then we felt the undercurrent of a cold wind. I realized now why Pasquier had insisted we should kit ourselves out with everything from cotton shirts to winterweight sweaters. It is this mixture of hot-and-cold, typical of the Saharan winter, that is the cause of so many of the inhabitants' pulmonary diseases. In Salah, like other central and southern oases, has a daily temperature range of at least 14 degrees Centigrade (25 degrees Fahrenheit). On the day when we arrived it was providing its unfortunate inhabitants with the equivalent of Algerian summer days and English winter nights.

We had no time to leave the airfield. It offered, in any case, its own curiosities. In the garden of a small clump of buildings where a hose jetted lazily, Pasquier told me to rub my fingers on the foliage of a tamarisk tree, then taste them. I did so. They were salt. The tree, he said, picks up salt from the ground through its roots and exudes it through its leaves. He pointed to a group of big trees just outside the compound. 'You see they are apparently standing on mounds? Now why should there be these particular mounds under these particular trees in this particularly flat stretch of sandy desert?' 'Tell me,' I said; but this would not do for the Professor. He bade me walk out of the compound to where a number of strange, steep-sided hillocks were grouped like pillars of salt. Which indeed is in a sense what they were. The skeleton of a branch, like the polished bones of a gesticulating human arm, projected from the top of one of them. Another was half-pillar, half-tree, with rags of foliage tossing from its top. I guessed that it was something to do with the root system. 'Precisely,' said the Professor. 'You tasted the salt. Now you see where the tree has exuded it in tiny sticky droplets, causing the sand to agglomerate, so that little by little a mound of hard rocklike salty sandy substance is built up round the tree until it is completely engulfed.' The nomads' name for such a hillock is *nebka*.

Leaving aside this aspect of its life and death, the common tamarisk of the wadis, *tamarix articulata*, is one of the extremely few species of trees and shrubs able to withstand the hostile environment of the open desert. Like the desert acacias, *raddiana* and *seyal*, it seems able to live through one rainless year after another. The small amount of moisture in the lower sub-soil is enough, apparently, to ensure survival. The botanist calls such plants xerophilous, meaning lovers of drought, and under pure desert conditions these alone are able to endure. In almost all desert perennial vegetation, nature has met the challenge of sun and drought by reducing the plants' organs of assimilation and transpiration, that is, the leaves, to the minimum. One way of doing this is to turn them into thorns, as in the acacias. Two years later I was to see many of these trees coated bright yellow by basking locusts in the full fig of their gregarious uniform.

The landing-ground at In Salah is out of sight of the oasis and in the airfield hutments it was possible to forget for a moment the bleak and terrible nature of its surroundings. Only when we were again in the air could we see the extent of the Sahara's stranglehold and the pitifully small area which men have been able to keep for their own. The town – for I suppose one must call it that – is strung out along the feet of a few low hills. Palm plantations and settlements provide evidence of water – and also of the desert's encroachment.

I photographed one which must have been half a mile long. It was the ghost of a village which had first been eaten by locusts, then by the desert itself. Sand, sifting through the trees, rose up into waves along the edges, making it look like an abandoned island in the sea. But the waves were dunes, forever on the move, forever piling a little higher, taking a little more from the cultivator and adding one more desolate fragment to the wilderness. Closer, almost under the plane, small geometrical patterns in the sand indicated former fields and vanished buildings. Beyond them, in the direction of higher ground, ran long straight lines of sandy bumps resembling giant molehills. Looking more intently, I could see that they had at some time been rings of soil at the tops of vents. Some seemed to be still in fair condition. Others had fallen in. Most were spaced at intervals of about fifty yards. Pasquier told me that they were the surface outlets of *foggara* shafts sunk into the sand for the purpose of making subterranean canals to bring water to a settlement.

An oasis is the result of water from an aquifer being forced up on encountering impermeable rock, thus creating a natural spring making organized life possible. Even 3,000 years ago, however, the Sahara's inhabitants needed more water than this if they were to maintain, let alone strengthen, their precarious hold. The *foggara* was their ingenious way of getting it. The first requirement was to locate the water table at its highest level. A shaft was then dug down to a point where the water could begin to accumulate, as it might be in a well. Once it flowed freely other shafts were sunk, continuing if necessary for many miles. As this work proceeded, or possibly in some cases at the end of it, the bottoms of the shafts were linked by a tunnel with an almost imperceptible incline – 1 in 1,000 or more – drawing off the contents of the aquifer in a steady, even flow until the tunnel broke surface close to the gardens to be irrigated.

Jews and Berbers are believed to have introduced the idea of the *foggara* from the eastern deserts well before the days of Christ; and they in their turn must have got it from still further east, probably from the Persian *qunat*, still very common in Iran. There is a possibility that the Desert Locust may also have played a part during plague periods, when the females, adept at seeking out moist ground for egg-laying, would in this way indicate to *foggara*-makers where to begin to dig for water. A FAO hydrologist, R. P. Ambroggi, has suggested* that a study of the locusts' behaviour could still be useful in locating groundwater. In the past the construction and maintenance of *foggara* were carried out by the forced labour of negroes in the service of the Tuareg. The result of the abolition of slavery is that many of the canals have fallen in. In the Touat, in the western Sahara, where the system was extensive, tunnels still in use are said to total more than 1,000 miles.

At In Salah there must have been a considerable mileage also. From the air they looked like symbols of a lost battle. Beyond them the sand found new obstacles against which to drift – masses of dark grey rock, smooth as a squeeze of toothpaste; cones, peaks and pillars of miniature mountains, sharp-ridged, flat-topped or pimpled and cut up by deep, steep-sided canyons winding out into sandy wadis. Mountains were growing higher; black crags falling sheer; a feeling of crescendo as the plane droned on over the last four hundred miles of its journey – the longest hop between any of the principal oases. Now, through the left-hand window, we saw the foothills of the Hoggars, now the north front of the range enfilading eastward, its huge peaks taking on the first tones of sunset glow, and next – for Tamanrasset lies at the south-west corner – the cliffs and crevices where the range ends below the peaks of Mont Tahat. This is the highest point in the Algerian Sahara – 9,730 feet. Only the Tibesti in north-west Chad surpasses it. The Hoggars are the last major landmarks before the desert begins its long slow descent into the Niger basin. They are also, because of the vast extent of their wadis and run-offs, keys to the possible behaviour of the Desert Locust in this region.

The locust-prospecting team awaited us at the airfield. Pasquier leaned over a French military map spread out over the bonnet of a Landrover while news of the most recent sightings of the insect was given to him. It was an attitude with which we were to become very familiar. We then drove into the last oasis.

* *Scientific American*, May 1966.

Chapter 6

Into the Emptiness

The name of the Sahara has been translated as The Great Emptiness, The Great Loneliness or The Great Nothingness. Tamanrasset is its outpost. It has the only inn on the thousand miles of track between In Salah and Agadez in Niger. It is also the starting-point and terminus for camel caravans and a staging-post for others, particularly those whose chief merchandise when going south is salt dug in the historic mines of the Hoggars.

A number of camels blocked the entrance to the town. We drove gingerly between and around them while their owners, Tuareg nomads, backed them hissing, snarling and foaming into a corner of a square beside a rustic Arc de Triomphe built of deep orange-coloured mud brick. Close to the gateway a notice announced the height of Tamanrasset as 1,390 metres (4,633 feet), which makes it easily the highest town in Algeria if not in the whole of the Sahara. After the warmth of the plain it struck pleasantly cool.

Beyond the gate the single main street narrowed between two rows of tamarisks and a number of busy little cave-like shops selling everything from Omo to camel cloths and saddles. Another, more formal square at the far end contained the prefecture and a barracks. Behind the sentries at the gate the Algerian flag took the evening wind. It billowed in semi-silhouette against the glow of mountain crags seeming to begin just outside the town. They were probably further, for the clarity of the desert atmosphere plays tricks with apparent distances. Just off the square lived Père de

Foucault, the French soldier-saint who was murdered by a pair of renegade Tuaregs, people he had spent much of his life trying to help. His hut is now a shrine.

This is the quiet end of Tamanrasset. We swung round in the square and pulled up near a seething mass of djellebas in the *suq*. The hotel L'Amenokal – named from the title of the former Tuareg rulers of the Ahaggar confederacy – is an unpretentious low white building standing in a small garden behind a wall. I filled in my card under a British European Airways poster luring me with composite pictures of a thatched cottage and a castle to 'Come to Britain'. The date in a corner was 1954. After a meal of couscous and chicken we discussed our plans.

The routes normally taken by the locust prospectors during their search form a web which at that time centred on Tamanrasset. Their surveys had taken them 300 miles southward and south-westward to two widely separated points (Tin Zaouaten and In Guezzam) on the Mali and Niger borders respectively; westward 200-300 miles to the gravel plains of the Tanezrouft; much the same distance north-west to the plain of Tidekelt; a good deal more than 300 miles north-eastward to the important Wadi Igharghar and its tributaries; and 500 miles eastward to Djanet. In addition they had criss-crossed between these radiating lines wherever conditions favourable to the locust were suspected. From Djanet to In Azoua and thence to In Guezzam, for instance, is 500 trackless miles. In all, the area surveyed during the breeding seasons is rather more than 300,000 square miles. Within this there are possibly a dozen human settlements. There is no lonelier land on the earth's surface, outside the polar regions. The Algerian team had just returned from the country south of the Hoggars. They reported seeing 'interesting' numbers of adult locusts, mostly in the solitarious phase, on the borders of Niger. Pasquier, who had originally intended to visit the Tanezrouft before going south, decided we should concentrate instead on the southward journey.

We left well before dawn but first had to drive north for an hour before finding a track diverging first westward then south-west to skirt the Hoggar's rocky outliers. It was still dark when, swallowing each other's dust, we turned away from the corrugated surface of the main track on to a much smaller, smoother one past the first

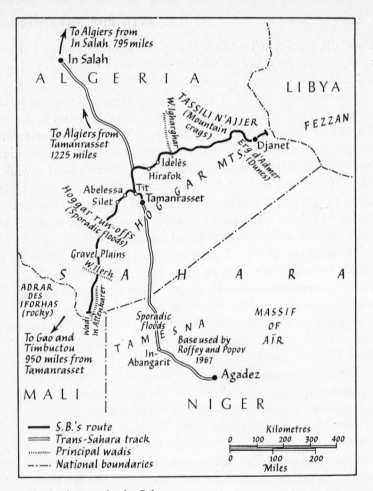

2 Author's route in the Sahara.

tiny sleeping oasis, called Tit. It is one of half a dozen minor ones looking toward Tamanrasset as their capital. With gestures worthy of Canute, Haratin women were sweeping the sand away from their doors as we passed through its neighbour Abelessa, a township once of great importance, for it was there that Tin-i-Nane, a celebrated Tuareg queen, was buried long ago. The early light, bathing a mountain valley, illuminated the mound of stones said to be her tomb, near the dry bed of a wadi which in her time had doubtless run with water. This marked the end, for a while, of our easy running.

7*a* The field study of locusts during recessional periods is essential in order to anticipate their future movements. The United Nations Development Programme provided the means for extensive surveys, which are coordinated by the Food and Agriculture Organization. Abdel Liman, an Algerian locust prospector, is one of the many whose job is to search among five million square miles of desert to which the Desert Locust retreats during recessions.

7*b* Professor Roger Pasquier, of Algiers, examines a non-swarming, or solitarious, specimen for signs of the colour change that may indicate the onset of the dangerous gregarious phase, when the insects multiply and swarm.

8 On this small patch of Saudi-Arabian sand, its size being indicated by the wheel track, there may be up to 100,000 hatchling locusts. The picture was made in the early morning, one to two hours after hatching, before they had turned black – most of the small black markings in the photograph are actually their shadows. Clifford Ashall, who took it, described the ground as seething with them. Egg pods were counted at densities of over 100 per square foot. The young hoppers numbered up to 1,000 per square foot at any one time, and hatching continued for several days. In Sudan in 1968 I saw similar egg-fields, shimmering with countless minuscule hoppers, extending several square miles. Within a few hours they would form into bands and march.

Beyond Abelessa, a long and weary stretch of lava boulders brought our speed down to ten miles an hour. Lurching and bumping, alternately climbing and falling, always with the bare rocky hills for company, we headed for Silet, the last oasis on the track, a poor place prefaced by a grey cindery plain with a fringe of ill-looking palms. A dry stream bed the width of a small field ditch wandered from a corner. We pulled up beside it close to a couple of mud huts. Bennounou Ag Rali, our guide, got down from the leading Landrover in which he had been riding with Pasquier. I never saw more of him than his eyes and the bridge of his straight nose, for he was of Tuareg blood, though not pure, and always wore the *teguelmoust*, the veil which is required dress for men of his tribe. Wound round his head and face, it was so arranged that he could slip his right hand – always the right – under a fold when eating, while with the other he held the cloth away, guarding his mouth from view. This is the complete reverse of Muslim practice in the north, where men expose their faces and only women go veiled. Bennounou's wives, who lived in the hut, were too shy to show themselves. They would, however, have been unveiled like all the Tuareg women and their Haratin neighbours, and this despite the fact that all are Muslims.

Many writers have purported to explain why Tuareg men always wear the veil. The Tuaregs' own explanation is that they do so because they always have! There are references to it by at least one eleventh-century Arab author, so it is a fair assumption that the practice is at least 1,000 years old and may well date from soon after the time when the Berbers, ancestors of the Tuaregs, first came south across the desert, having been driven away from the coast by the Arabs. If this is so the reason for the *teguelmoust* is probably at core a practical one, for the Sahara was then already in an advanced state of desiccation with hot dry winds seeking every gap in tent or clothing. It would be natural for men riding the desert in all seasons to devise some form of protection for nose and throat. Catarrhal conditions, caused by fine dust acting as an irritant, are common among desert-dwelling Arabs who do not use the veil. In Saudi Arabia I have heard a most distinguished emir hawking repeatedly while conducting a village council meeting. The *teguelmoust* seems a sensible form of shield because

F

the cloth not only keeps out the sand but forms an envelope in which there is always a little moist air from the mouth serving to protect throat and nostrils from excessive dryness. But with the passing of the centuries, evidently, the wearing of it has become, in addition, a social ritual no self-respecting Tuareg would forgo, so that when Bennounou said that he wore it for decency's sake he meant precisely that.

From this point onwards our safety was very much in Bennounou's hands. It would be Bennounou, moreover, whose knowledge of the Sahara would be the chief aid in our quest. Much of it was, of course, inherited, for, like every other man in the oasis where he was born, he had learned the ways of the desert at the age of eight or earlier, when he first travelled with his father's and grandfather's caravans. At seventeen he joined the French Army Camel Corps, with whom he spent the next twenty-four years. His service as a guide followed. At the time of our journey he was forty-three and had spent all his years in the desert. To Bennounou the Sahara had finite edges represented by the limits of his experience. Beyond a range of about 500 miles the world ceased to exist for him. Within it he knew everything and every living person. By looking at hard ground where to us no track was visible he could tell when the last caravan had passed, how many camels it contained, whether they were lightly, modestly or heavily laden and usually – because the loading of camels tends to have individual variations – who were the owners. Tucked away in his memory was the date and duration of every rainfall within his own lifetime and those of his father and grandfather. He had a photographic knowledge of every furrow of the torrential run-offs spreading far out from the mountain watersheds, the whereabouts and condition of the few wadi trees and, most important, the nature and amount of the ephemeral vegetation likely to spring up after any recent rainfall in any given place.

Physically, Bennounou was like most other Tuaregs, tall and thin, narrow-hipped, flat-chested and long in arm and leg – a whipcord of a man. His nose, what could be seen of it, was straight and thin. His eyes were dark and seemed perhaps the more expressive because of the concealment of his other features. Pasquier believed that he had a touch of Haratin blood. This would have been likely. The Haratin are negroid people who entered the Sahara from the south a considerable time ago, either as immigrants or slaves taken by the Tuaregs in order to cultivate the oases. Until the coming of independence and the rise of the black republics they remained semi-vassals, receiving a fifth of the garden's yield. In theory, this state of affairs is now ended. In practice the relationship has become an uneasy one in which no one is certain who is top dog. The Anti-Slavery Society of London has stated that there are still 20,000 slaves in the Tamanrasset area. The Algerian government denies there are any slaves at all. Since their writ is only nominal in the very remote oases the truth is likely to be half-way between; but slavery in the Sahara is very unlikely to resemble a state that Wilberforce would have recognized.

About 450,000 Tuareg are today scattered on either side of the no-man's-land of Algeria's borders with Mali and Niger. Since the latter received their independence there have been reports of a considerable flow of Tuareg refugees northward into Algeria. If they are true, no one at the official level in any of the three countries seems to wish to know about it, and it is impossible to get reliable figures. The Tuareg are indeed paying the price of their own history. As cameleer-warriors their prosperity has been geared for the past two thousand years to the caravan trade. This is now declining and with its extinction their *raison d'être* will cease to exist. Every aspect of the only life they know runs counter to the logic of progress. Far away in Europe, concerned and kindly people are devoting a good deal of thought to the possibility of settlement schemes; but where? Disregarding the fact that trying to turn a caravan drover into a farmer would be about as easy as harnessing a racehorse to the plough, there is at present no available suitable land, certainly not in Algeria and I suspect not in either of the other two affected countries, where the indigenous bush farmer already has a desperate struggle for existence. If and when science brings more water to the desert, then the picture may be different, but I fear the tribe of Bennounou will have decayed and vanished, like the walls of his village, long before then.

It was pleasing, therefore, to discover that Bennounou had been able to preserve a little official dignity, for he now revealed himself to be not only our guide but the mayor or at least the head man of Silet. With diffident pride he showed us his adobe house occupying the best corner, close to the streamlet. At a right angle to it a mud wall protected his garden from goats. His nephew, a handsome lad of fifteen, had grown some vegetables and a little corn.

Together with a few hens, this was our guide's whole estate, his buttress against the wilderness lapping at every wall. On it he supported two wives, a sister and their children. The few other villagers were no better off. The oasis had the air of a forgotten outpost, long ago overrun. About half a mile from Bennounou's house, on the edge of a tattered palm plantation, the roofs of an abandoned Foreign Legion building sagged toward the ground. The remains of what had been a formal forecourt opened through a gateway into an interior court with a well where a couple of palms stood sentry. Pasquier pointed up to them. Up to a height of about twenty feet their trunks were perfectly regular, but were then pinched in, showing where locusts had eaten away the heads during the last plague. The new thin growth would hardly have fed a single soldier, had any been left to gather it.

During the last war, when the outpost had been manned, French troops had, without much difficulty, scraped out a small dead level space where a light plane could land among the grey lava. In a more fertile land it would long since have disappeared, choked by weeds and trees, but in this place nothing grows, not even the thin aristida grass which now and again – at intervals of perhaps fifty miles or so – spreads an unexpected evanescent silky silvery sheen, wonderful to see, where a little humidity has lingered in the run-offs of past torrents. So the little strip remains, an historic fossil which, as Hadj Benzaza, taking a turn around it in the Landrover, said wryly, might yet be useful in the event of rescue ever being needed. This, of course, would require a radio link between the vehicles and the sub-prefecture at Tamanrasset, who could then ask for help from Fort Flatters, an oil company air station 450 miles north-east. But at this stage there was no radio link – a lack, I am glad to say, which has since been repaired as part of the UN project.

It was now about 9 a.m. and we had driven a hundred miles.

The sun, scouring the air strip, had resumed its cleansing job on the body of a dead camel lying in the middle. The heat parched and cauterized the flaps of flesh and skin, purifying them as it does every other dead thing in the desert. At the edge of the lava, the land level dropped by a foot, forming a ridge between a grey land and a red one made of particles of glowing gravel over which we sped in echelon at forty miles an hour. This was the beginning of the regs, vast plains, luminous and lonely, spread like cloths between areas of dunes and rock. As we headed into this nothingness the increasing heat turned the landscape upside-down, dissolving the horizon into mirages so that we seemed to be driving on moving islands amid molten seas of refracted light. As fast as we drove we could never reach the edge, but remained always in the middle. From our position in the second Landrover, looking ahead at the leading one, it seemed as if Pasquier, Abdel and Bennounou were forever on the point of vanishing. To the party behind, we also were reduced to a single pale shape, an unstable landmark faintly silhouetted against the shimmering vacancy where gravel and mirage mixed.

If at Tamanrasset I had had the feeling of being temporarily at the end of civilization, it was on this stage south of Silet that I began to realize, in a purely emotional way, the true immensity of the Sahara. It was to be measured, it seemed, not by distances or square mileage or the time taken in moving from one waste of rock or gravel or sand to another, or even from one waterhole to the next, for there is a point at which such measures cease to have any sort of bearing on personal experience. One can say that the total desert area of the Sahara is four million square miles, but it is difficult to visualize such a space. Somewhere south of us the maps showed a national boundary line running 800 miles dead straight between two points each in the middle of nowhere. In theory it divides the western Sahara into two neat halves. But no one has ever travelled it, nor would nor could do so, for there is nothing on it but sands, unutterably lonely gravels and the wandering con-tours of a single wadi. It is this sort of line over which nations quarrel but which the Sahara shows for what it is, a cartographic fiction serving bureaucratic absurdity. The only measure which had any validity at all was the one we carried with us in the shifting relationship between our eyes and minds on the one part and the contracting or receding horizon on the other.

For long spells the track was visible only to Bennounou. The incidence of rock drawings in the Hoggar and other mountains showing four-horse chariots of the type described by Herodotus as being used in the eastern-central desert, has led some writers to conclude that this was part of a pre-Roman route from Tripoli to Gao on the Niger, but except for a few stone cairns piled up for the guidance of the French Foreign Legion so little is to be seen on the ground now that it might never have existed. Today only the wheels of the locust survey vehicles pass this way, and wind and sand erase their marks almost as soon as they are made.

Thus we proceeded on our separate but overlapping islands, for hour after hour, yet it cannot be said that the journey was monotonous, for as the mirages ebbed and flowed around us so, almost imperceptibly, they continuously changed their shapes, now creating phantom headlands, bays and inlets and next extinguishing them – all alterations depending on slight variations of the gravel surface and on the layers of the hot air above. Now and again the image of a rock would be duplicated, casting a watery reflection. Like a sun-crazed creature in a comic cartoon I would find myself imagining boats and even water-edge cafés, so alike were these illusions to the real thing. Lulled by the throb of the motors, we seemed to be moving through a world in which actuality and illusion had been temporarily transposed.

But there was always Pasquier to bring us back to the purpose of our quest. From the moment of our leaving Algiers he had evidently decided that in uninhabited country there is no point in sophisticated superficialities. His ancient grey suit was the one he would sleep in; his beret and his boots, honoured companions of many a Saharan journey, had reached that point of disintegration where clothes become part of their owner's persona and, decaying no further, will faithfully protect him for the rest of his days. His beautiful iron-grey hair, which looked so smart when combed back and brushed for our visit to the sub-prefect at Tamanrasset, stood up like a windblown aura. Stubble, silvering his jaws and chin, framed the beginnings of a fine new mellow tan. His eyes, always ready to kindle like those of the good teacher he was, had a brighter gleam. We awoke from our moving world of illusion and dream to see that he had stopped his Landrover and dismounted and was beckoning us to get out. With shoulders hunched, he peered at some clumps of desiccated herbage.

'*Hyoscyamus falezlez*,' he pronounced, 'very good for locusts!' 'The Tuaregs call it falélé, but as we have such good Moslems here' – he glanced quizzically at the driver-prospectors, Abdel, Fuad and Mahomed, sharing a can of beer – 'we'll use the Arab word, bettina.' Taking a stick, he walked slowly over the ground, pointing to this plant and that, making the drivers recite their names. The selection was not very big – morkba (*panicum turgidum*), a tall grass with seeds used in hard times to eke out millet; another kind of grass (*chloris barbata*) called kerdenellagh by Bennounou; and gartoufa, a low-growing plant from whose leaves he was able to produce a very fair substitute for tea.

The sight of a few weary-looking acacias, rather than any perceptible change in the desert's level, confirmed that we had reached the edge of a wadi, one of several fanning out from the mountains, now about a hundred miles away in the north. Bennounou agreed with Pasquier that it must be a branch of the great Wadi Tamanrasset whose intermittent waters wander far westward to peter out in the Tanezrouft. Popov, crossing this region during his survey, found no blade of anything in five hundred miles. Yet, as he has pointed out, there is never a year without rainfall somewhere in the Sahara. What makes the desert habitable for men, some plants and locusts is not occasional light and scattered rain, evaporating as fast as it falls, but heavy and concentrated downpours in the highlands whose force and volume send the waters coursing far out through more or less permanent systems of wadis. Although these may be superficially dry for years on end during which no rain falls, they are nevertheless able to retain a slight degree of moisture.

Storms such as these are associated chiefly with the climatic feature known to meteorologists as the Inter-Tropical Convergence Zone. This is the band, stretching right round the world, in which opposing hot winds from either side of the Thermal Equator meet. At ground level they are one of the causes of the great deserts, but at the point of meeting, where they rise, they sometimes give off rain in cataclysmic storms. Then, woe betide anyone who has camped in any of the areas of the most intensive run-off. Amid Valkyrian thunder-rolls and lightning flashes the storms break over high ground where moist air flowing upward over the slopes has produced a build-up of condensation. Almost between one moment and the next this is discharged in violent torrents over-whelming rocks and hollows and surging out through every dis-coverable gap into erstwhile arid wadis cut in the desert by previous storms. Down these it rushes in wall-like waves, dividing and sub-dividing until nothing is left but a few trickles. By then the flood will have travelled perhaps 200 miles or more and along its path there will be a number of areas sufficiently moisture-retentive for some plant and tree life to gain and main-tain its hold. Sheet flooding of the occasional depressions to be found in regs may produce similar results. In either case it is only the most highly adaptable species which are able to withstand the hostile conditions: years without further rainfall, temperatures ranging from burning hot to bitterly cold, and relentless winds.

One of the continual surprises in the Sahara is that, despite to all appearances every plant being shrivelled and every tree and shrub apparently on the brink between life and death, there are often traces of moisture a few inches down which keep the perennial vegetation in its state of suspended animation, ready to burst forth afresh when the rains come. Lying on the ground dusted over with sand there will also be the seeds of annual plants, and these too are patiently awaiting their opportunity to germinate. In general, so great is the desert heat that any rainfall of less than about a third of an inch (ten millimetres) evaporates immediately unless coming on top of previous rain. To produce good vege-tation a fall of at least an inch is necessary – less would only cause an ephemeral flush of germination and such growth as occurred would quickly wither and die. The characteristic desert plants avoid this fate by a process of adaptation as old as the desert itself. Between rainfall and ensuing drought annuals are able to mature, flower and cast their ripened seed. Perennials awaken from their long sleep and also flower, as do certain trees. All this takes place in a very short time. In the case of annuals the cycle of life may be completed within as little as three weeks. A favourable feeding plant for the Desert Locust, the tall blue-flowered schouwia, on the other hand, seems able to spread its period of development, some seeds germinating quickly after the rains while the rest wait two or three months.

Whether by divine design or not, these are precisely the conditions the locust requires for its own survival. It, too, needs sands for its egg-laying. Like the plant seeds it, too, needs soil moisture equivalent to about an inch of rainfall for the eggs to germinate successfully. It, too, must accommodate the most important part of its life cycle to the brief spell in which the wadi springs to life. But unlike the plants it can seldom wait to achieve full maturity but must eventually move on to seek new pastures. Just as nomadism is an essential part of man's defensive mechanism against the hostility of the desert, so it is with the Desert Locust. Over aeons of time it has adapted itself to these conditions by migrating from one area of rainfall to another. To be precise, there are areas where rain has already ensured its requirements for survival. If the rain fails over a wide area, the locust population of that particular area may die out. This may account for the fact that countries which have at times known the full force of the plague have at others had not a single locust within their borders. But always in some desert wadi somewhere in these limitless wastes there is bound to be a Desert Locust population only awaiting the needful conditions in order to expand. Whether it is these unknown solitarious populations which are primarily responsible for the carry-over from one plague to the next, or whether somewhere in these vast spaces there are always locusts in the dangerous gregarious phase is a question entomologists are still discussing.

Meantime, it appeared that someone had been ahead of us. Slobbering and frothing at the lips and uttering angry hisses, a white male dromedary circled the cars while his females grazed peaceably in the mid-distance. Abdel aimed a stone at him without much effect. Then, picking up the long-handled nets which had been stowed behind the car seats, we fanned out over the flat wadi bed, which was here about half a mile wide. I took my place alongside Abdel. He said: 'We saw locusts here when we passed a month ago. It was greener then and I think they will have gone, but the Professor wants to be sure.' He walked about, searching between the clumps for locust droppings or the remnants of any insects which might have fallen a prey to birds. Now and again he pushed the stick handle gently into a patch of leaves, stirring them and watching for movement. There was none. The dromedary had decided to accompany us and Abdel threw another stone. He moved off a yard or two with teeth bared and lips dripping saliva, but continued to keep pace between us and his wives. Bennounou meantime had located the owner, a neighbour from Silet. He had ridden a hundred miles in three days in order to pasture his herd and had seen no locusts where we were searching.

We got back into the cars, drove along the wadi edge and then, at a point where it narrowed between banks some three feet high, lurched down into the middle of it, driving through a mass of small flowers of a faded blue borne on brittle stalks about five feet high and filling the wadi from side to side. Beyond the bonnet, a few minutes later, a pale-bodied creature about two inches long leapt away forward and, planing in a long iridescent rising and descending path, vanished into the blue of the schouwia fifteen yards away. We had flushed our first locust. Within the next ten minutes we had bagged three. This hardly suggested anything so menacing as a swarm. Holding and turning them so that the sunlight played over their bodies and through their wings at all angles, Pasquier pronounced them to be young solitaries. He handed them to Fuad, who popped them into a killing bottle. Spreading their prepared record forms on the hot bonnets of their vehicles, the three Algerians noted the date, the time, the map reference and the plants on which they had been caught. Fuad, taking out a starting handle, poked it into the soil to a depth of about ten inches and scooping out a handful showed it to Pasquier. He pronounced the humidity to be very slight – an indication that although here and there in the wadi locusts might be breeding, the conditions were not really ideal.

It was a good, if not exciting, beginning. Contented, we lunched off tinned sardines and raw onions near the shade of the Landrovers on the open ground beyond the wadi, then we embarked on the crossing of what Pasquier said was a part of the Great Reg Ténéré. Spreading down into Niger where it joins up with a belt of fixed sand extending right across Africa, it is a kind of land ocean made of various grades of pebbles and gravel, and I wish there were something more dramatic to say of it than that it provided us with the best motoring of our journey. Weathered by millennia of erosion into a high-speed track in which it would have been possible to drive in any direction at forty miles an hour it presented only the question: which direction? We put our trust in Bennounou. Happily unaware that in previous journeys with our drivers he had kept to a cairned route, I relaxed in such comfort as one can contrive when sitting three together on the two front seats of a Landrover while he veered east of it, taking a short-cut clearly marked out in his own mind (but no one else's) by the position of the sun, the nature of the gravel, the occasional 'steps' where it changed level, a few faint curves of the surface and an occasional camel dropping, all helped out no doubt by ancestral memories.

Driving almost up to the limit of our vehicles we watched the sun slide down the sky until toward evening it illuminated an inverted arc of cirrous cloud, flushing it pink across the whole western horizon. The purpose of our guide's short-cut now became obvious. It was not to save time or distance but to bring us to a suitable camping place which he had noted when passing by camel many years before. Pasquier and the drivers eyed approvingly a group of decayed tamarisks, one of them dead. While we began unloading, Bennounou started off toward the dead tree, intending to fetch firewood. A cry attracted us over to him. Bent over a patch of bare soil between clumps of panicum, he unearthed a Tuareg skull. With half its teeth still intact, it grinned up at us out of the sand, where it had evidently been washed from a higher level. The name of the wadi was Wadi Ilherh. We decided to call it Skull River.

While the fire climbed and the drivers worked and cursed, erecting a tent with frame apparently constructed on the principles of the Eiffel Tower and of such complication as has probably never before been seen in the desert, Bennounou, no mechanic, made tea. He tipped water, without spilling any, from the jerrycan into a tin pot which he balanced on two stout pieces of flaming timber, and then, when it was boiling, poured it on to the dried gartoufa in a tin teapot decorated with painted flowers. This was a

very special possession. After using it he scoured it with sand, the
cups likewise, rubbed it lovingly until it gleamed and, wrapping it
in the same cloth, put it carefully back behind his seat in the Landrover. The tea, strongly sweetened while still in the pot, tasted
good. Fuad next took over the fire, refilling the cookpot with equal
care to spill nothing and then pouring in a quantity of short pasta
of unique flavourlessness which he boiled for forty minutes. We
ate this silently, yearning for the salt which had accidentally been
omitted when packing the provisions at Tamanrasset. Meantime
Fuad had also been baking bread. It had the quality which I
imagine lava has when it begins to solidify. We ate it with lumps
of corned beef and much enjoyment, aided by another can of
lukewarm beer.

Some of us washed before the meal and some after, using half a
cupful of water poured over two pairs of hands, so that those below
caught any spilled by those above. Bennounou combined his
washing with the tea preparation. It did not seem to affect the
taste and he smelled better than any of us.

Pasquier, having supervised the tower-building with ironic
comment on those who had designed and selected it, announced
that he proposed to roll himself in a garment which had once been
an overcoat and would sleep on the sand. But first, warmed and
full, we must consider the events of the day and learn from what
we had seen. 'I shall speak in English,' he said. 'It is an absurdly
illogical language which I have not used for years, although I read
it well. I shall thus be compelled to present my ideas clearly and
simply and you will be able to understand them. We shall have an
intellectual *entente*.' It was to be the first of a series of lectures,
unforgettable equally for their style and for their setting of flicker-
ing embers and the circle of night beyond them and the stars
above. These nightly talks are the basis of much in the succeeding
(and indeed in the foregoing) pages.

Chapter 7

Wadi In Attenkarer

A characteristic of waking up in the desert is that what looked very snug the night before, when lit by the camp fire, can look pretty bleak at dawn. So it was now. The wadi, or at any rate that section in which we found ourselves, was revealed to be shallow and narrow and its vegetation so completely dried out that it was impossible to think that it could ever spring to life again. Beyond this thin sliver of dead river the gravel reg, hot red in the early light, stretched southward, vast as ever; but now at least there was the pleasure of seeing clumps of low mountains rising here and there like islands from the level surface. The sun rose, beguiling our morning splendidly with a series of 'architectural' mirages quite unlike any we had seen before.

The first appeared in the form of a small rectangular arch cut through the mountain. Next, a rectangle would be bitten from the top, gradually extending until the whole top level of the mountain disappeared into apparently clear sky, leaving a thin, black layer bridging a line of arches below. At one moment there would be a viaduct high in the sky and apparently many miles long, and at the next the whole deck would vanish, leaving only isolated stacks of rock. Alternatively, the bottom would be wiped out so that the range became a chain of unsupported notches. In the last stage but one, most of the mountain had taken the form of a gigantic diving-board, supported on one end only by a stump of rock. Finally, practically the whole range dissolved.

At 9.20 a.m., beyond a colossal field of large boulders, wind-carved and sun-baked into the semblance of golden heads and torsos, we again left the cairned route and branched leftward over open desert and firm sand to a gap between low granite hillocks. It opened into a long river of sand between low dune-like formations where the sand had been blown over rock. This was the beginning of the Wadi In Sarfi, and we stuck to it for an hour until, crossing some blue rock and climbing a short rise, we came face to face with a line of high crags half buried in bands of soft yellow drifted sand. The beauty of this spot is very hard to convey for it really consisted only of the jet black rock, the brilliant sand and the dark shapes of a few acacia trees suggesting this had once been one of those green spots which Tuaregs quote their great-grandfathers as remembering before the desert had advanced so far to the south. There was also, farther up, a small stone circular tomb, probably of the Stone Age. It lay enigmatically on the slope of a fine valley through which we slowly threaded upward, still between the crags, to a grey moraine upland. Pasquier confirmed that this was part of the Tassili du Ahaggar, eroded remnants of the range whose peaks above Tamanrasset lay at least three hundred miles to the north-east.

Since leaving Silet a day and a half earlier we had seen no sign of water and met no one. Now, rounding a bend, we came suddenly on both. The In Tedeini Well – the Well of the Ticks – is 200 miles from anywhere on a nomad track invisible to the untutored eye but clear enough to Bennounou. Ticks thrive wherever there are camels. A dozen camels crowded at the well, teeth bared and dripping while a cropheaded boy hung on to their lead ropes, hauling with infant authority against their stretching necks while his father, swathed to the eyes, hauled bucket after bucket of brackish water to fill innumerable goatskins. The man told Bennounou that they had come up from Niamey in Niger, thirty days march to the south, and were headed for Tamanrasset. Bennounou advised us against drinking, adding cheerfully that sometimes nomads fell in the wells and died. He knew a woman who had done so, tripped by a camel rope. He then interrogated the merchant about locusts. The man had seen none but mentioned that he and his son had passed a Negro nomad family with goats

and cattle some hours before, in a wadi where there was pasture.
Pasquier listened with interest and asked Bennounou to alter our
route to find them. We came upon them in the early afternoon
camped in a black goatskin tent amid desiccated vegetation
surrounded by a crypto-moonscape of red gravel and grey shale.
Half a dozen cows and about twenty goats signalled our arrival by
a wild stampede. A man, his two wives and a multitude of naked
children peered at us from beside a fire. They hung back shyly
and were unable or unwilling to answer Pasquier's questions, put
through Bennounou. Where had they watered? Had they been
to the well? We had seen no droppings. Had they seen locusts?
Where had they come from? The man gestured south-eastward
in the direction of the Niger border, where there is good water
near the guard post at In Guezzam, some 150 miles off. We were
ourselves, at this point, well into the south of the Sahara and not
far short of the Tamesna, where there had been autumn rain.
Pasquier conjectured that the rains could have spread northward,
improving the vegetation and perhaps even leaving a little
standing water here and there. Thus encouraged, the nomads had
moved in their track, no doubt picking up information from an
occasional caravan, until they had at last arrived at this dreary
haven. If this were so, it suggested that somewhere in the region
there must still be conditions favourable to locusts. We arranged
to buy a goat for our dinner on our return and with a new member
literally attached to the party in the form of an aggravated
twelve-inch-long palm lizard which fastened its jaws in the
Professor's thumb while he searched for locust traces in a clump of
kerdenellagh grass, we pressed on southwards, around the end of
a line of low black hills resembling industrial slag heaps at their
grimmest, on to another big reg, graced with the skimming shapes
of alarmed gazelles. One group of these exquisite creatures crossing
in front of us must have been moving at least at fifty miles an hour.
They paced us easily for half a mile before veering off to the screen
of a patch of acacias in a long wide wadi. A few minutes later we
struck the wadi at a different point and, following it down,
found ourselves amid such dense schouwia that soon the cars
were lost to sight of each other as they lurched and swayed
through it. Through the windscreen we saw the repeated flash

of iridescent insect wings. Here were locusts, and in some abundance.

The Wadi In Attenkarer is some 250 miles long and stretches across the border from the Tassili du Ahaggar in Algeria, from whence come its periodic floods, to the western Tamesna in Mali, where it is associated with a complex of wadis once serving as tributaries to the Niger. It is halfway between a pair of tracks two hundred miles apart and is unlikely to be visited by anyone except the anti-locust teams of Mali or Niger. Pasquier was pretty sure that no Europeans had ever been there, not even the colonial French. Our team had seen signs of a minor locust build-up when passing this way a month earlier. This is a classic situation for the Desert Locust: a breeding place not too far, in locust terms, from the desert fringes, adequate moisture ensuring plentiful vegetation and so far off the beaten track that observation is unlikely.

In the past locust reporting has depended on desert dwellers and travellers who have not only been restricted in their range but in their knowledge of the insect, so that the critical stages preparatory to a plague have been able to develop before counter-measures were possible. Moreover, these stages are not confined to a single district. In the present state of locust information it is generally accepted that three successful generations are necessary before an outbreak occurs, and between each generation a swarm may fly many hundreds of miles. By the time the control team has arrived (always assuming there is one within a thousand miles), the locusts may have bred and departed. If, therefore, one can locate the first generation there is a better chance of control.

The accepted method of estimating the numbers of locusts in a given square kilometre of vegetation is for members of the surveying team to walk through it in parallel rows about ten metres apart, stirring up the plants with a long stick. The numbers of locusts thus disturbed are then multiplied by 150, and this sum is divided by the number of prospectors multiplied by the number of kilometres surveyed.

Which is all very well in theory but is another matter entirely when wading under a burning sun through schouwia as high as one's head. Our drivers were hard put to see each other as they pushed their way, nets in hand, down the wadi. Nevertheless Hadj, standing on the bonnet of a Landrover, made what he thought to be a reasonably accurate count of the insects flushed in a fairly small section. This procedure was repeated elsewhere and again among the more open plants at the edge of the territory. Pasquier reckoned that there might be a million or more locusts in some ten miles of the wadi on both sides of the border, which, for reasons of international protocol, we were unable to cross, although the frontier was at this point invisible, unmanned and in all practical respects meaningless.* Meantime the prospectors were catching locusts to such good effect that the killing bottles were filled several times over. Pasquier, beaming and rubbing his hands, would have liked to have camped overnight to study the position more closely. But we were in any case now in a quandary. We had taken longer than expected to reach our goal. We carried provisions and water for only four days. We had no radio to notify our whereabouts in case of trouble. We were a hundred miles off the nearest track and we must not merely retrace our way to Tamanrasset but go several hundred miles further – as far east, in fact, as we were now south. The sun, moreover, was dipping and we were far from a source of camp fire fuel.

There is a feeling of frustration in being compelled to return on one's tracks; yet this soon vanished, for with the sun behind us a wonderful evening light bathed rocks and hills in a brilliant tangerine glow. The evening also brought out many more gazelles. One fine male stood steadfastly on a rock watching as we passed within a couple of hundred yards – it would not have done so had this been a main track used by men with guns; the Algerian locust prospectors carry none. Far ahead a couple of bustards, black spots in the sky, circled looking for life in a seemingly dead wadi. They were still searching as we passed. In the wadi of the nomads their animals again panicked on hearing our motors. Hadj and the owner chased the goats, singling one out. They bundled it

vultures [handwritten marginal note]

* The Prefect of the Eastern Oases, at Ouargla, told us later that had we informed him of our movements he would have arranged the necessary authority with his opposite number in Mali. The situation may be better now. (See Chapter 15, p. 222.)

G

whimpering into the back of the car, from which pitiful bleats came every time we lurched.

At the In Tedeini well it was already dark but Bennounou decided we must press on, and we fled through the night, scarcely ever altering our speed for an hour and a half as his incredible memory and sense of direction, aided by changes in the surface level, which to us seemed imperceptible but to him were as large and clear as signposts, steered us straight to the Wadi In Selfini and the spot where years before he had observed a dying tree. I had imagined that for him to have found it thus unerringly our camp place must be fairly large. We awoke next morning to find that it consisted of three tamarisks between two small hills. It also contained the broken remains of a German car whose American occupants had been lost ten years previously.

We had been going some hours, re-crossing the great plain, when faint specks in the extreme distance began to take shape as a file of fifty or sixty camels head to tail. We pulled up a quarter of a mile away and waited. The caravan came on steadily, passing a hundred yards to the left, for no caravan stops for a chance encounter. Camels fight, entangle their linked ropes and displace their loads at the slightest opportunity. The Tuareg leader, on foot, marched ahead, scarcely even turning to look at us while eight of his companions, all men and of all ages from bearded elders to laughing boys, detached themselves to greet us. They were evidently well known to Bennounou, whom they welcomed warmly yet with a kind of formal politeness, accompanied by much hand-shaking all round. Where were they bound for? To Timbuktu, twenty days' march away. They accepted cigarettes and after a little conversation ran to catch up with the camels.

We camped at Silet in the old fort and dined at Bennounou's house. As mayor of Silet, he wished to show his hospitality and laid out carpets on the sand floor of the guest room in one corner of the courtyard. A bright eye, belonging to a pretty ten-year-old, his daughter, flashed for a second round the doorway, excitedly taking in the forbidden scene. Composed cross-legged in a circle, we began with the ritual tea-drinking. Then came the couscous and perhaps it was because of the charming air of hospitality, the true sense of occasion inspired by our guide, that it seemed to me to be the best couscous I had tasted in Algeria. With it came piles of unleavened bread so that by opening it we were able to take up handfuls of food without the need for spoons. The remnants of the goat appeared as a side dish. We finished with oranges and then again had tea.

After 1,200 kilometres of practically empty desert Tamanrasset appeared as a positive metropolis, but Pasquier would not allow us to stay longer than needed to re-fuel. This meant treble-tracking, into the oasis by the way we had left it and then back again to the eastward-branching 'Piste du Hoggar', where a triangular concrete block with fading painted names and kilometrage figures on its side speaks better than words of the loneliness of the central and southern oases: southward, Tamanrasset 119, Agadez 1,019; northward, In Salah 581, El Golea 980, Algiers 1,930; eastward the only three inhabited places on our route were Hirafok 70, Ideles 100 and Djanet 616.

Pasquier's promise that this part of the journey would be completely different from anything we had seen before was confirmed by the continued presence of the Hoggars, their great peaks and clustered columns of basalt enfilading eastward while the setting sun turned them first to glowing pink sugar candy and then, as the shadows grew, to blue. For about seventy miles, from west to east, they maintain a magnificent front whose summits are never less than 7,500 feet: then they begin to crumble into a belt of fantastic chaos, part of the Hoggar Tassili.

Prehistoric paintings of dancing figures and wild animals in the caves high up on the slopes are clear evidence that this part at least was once a well-inhabited land with woods and rushing rivers where a huntsman could find all he needed to fill his belly.

Today the route from Hirafok to Djanet could hardly support a locust. It lies across more than 300 miles of a shattered, riven landscape and has natural vegetation in only two places, where the branches of a wadi cross it. We slept near Hirafok and before starting out to cross this wilderness called at Ideles, the last oasis on the threshold.

It has three or four hundred inhabitants. Some of their elders came to meet us as we drove over an open stony slope with the first houses of the oasis silhouetted on the eastern edge against the distant background of a cone-shaped mountain. Tall, good-looking, reserved, they at first seemed a little suspicious of our visit. Then one came running forward, crying gladly, 'Bennounou!' Another, a well-built Negro in a snow-white robe, clutched a packet of Omo under his arm. . . . A Muslim priest and a visiting official, a dapper young man in a djelleba, with the smoothness of metropolitan politics in his smile, joined the welcoming party. Pasquier, Hadj and Bennounou went off with them to explain our mission while I wandered around the village. The source of its existence is a clear stream issuing out of a gulley about the width of an English field ditch. Climbing past two young men sitting beside it I found the outlet of a *foggara*, and skimming around it, a few tiny fish. Downstream, in a wide flat space, beyond the gulley, scraggy hens and children with eyes distorted by disease played in and out of the water. A little girl crouched by it washing clothes.

During the French occupation Ideles acquired a miniature formal square with mud walls and a garden plot in the centre. One can imagine a handful of troops assembling there for a flag ceremony, the blare of their bugles and the crash of their grounding rifles echoing through a few hundred yards of their tiny civilization to falter and fade in the huge, unechoing, unforgiving wilderness surrounding it. No one uses the parade ground now. No one tends its garden. Outside its walls the houses are strewn haphazard over the slope toward the stream. Some are built of mud. Others are zeribahs, grass huts made of elephant reed, which is found in most of the oases. Ideles also grows tomatoes. The crop is pulped in a small handmill and carried by the occasional trader going to Tamanrasset. A notice painted over the door says that the mill is dedicated to the revolutionary people of Ideles. A woman pounded millet with a stone pestle and mortar in a hut nearby. In another hut a leather worker made camel saddles while a woman wove their cloths. I heard the sound of a hammer and found a smithy. The smith sat cross-legged on the earth floor making a beautiful and intricate lock, in which the key was represented by a stylized male figure and the lock by a female. These locks are highly valued as far away as Timbuktu, where they are sold for a good price. I wanted to buy one, but Pasquier had been

acting as my banker ever since my failure to cash travellers'
cheques at Algiers and said we would probably need our money
at Djanet. I was sorry, for the smiths of the Hoggars represent one
of the Sahara's oldest mysteries. No one knows whence they came
and they are neither of Tuareg nor of Haratin stock.

Such vegetation as there is in the outskirts of Ideles consists
largely of Christ's Thorns, shrubs of the genus *Zizyphus Spini-
Christi* from which Jesus' crown is said to have been made. That
plant life can get a hold even in the stony uplands was evident
where a few boulders had been broken up to repair the track. In
most of the pits so formed one or other of half a dozen different
species had spread its roots searching for moisture probably
deposited by dew. Beyond the hill there was yet more evidence of
the Sahara's pluvial past and of the ability of even quite large
trees to adapt themselves to the arid present, given a rare heavy
rainfall. The wadi into which we descended was wider and deeper
then any we had seen before, with well-defined banks like those of
a true river. Great boulders, some as high as cottages, littered the
sides of it or stood where they had been tumbled along the wadi
bed. Beside it groups of a dozen or so well-grown tamarisks shaded
the edges of a river of rippled sand. It was quite dry. At the further
side the track rose gradually over rockstrewn slopes, then fell again
to a similar wadi perhaps even better defined.

Both these valleys are the upland branches of a fossil river, the
Wadi Irgharghar, which had at one time watered the whole of
the eastern Algerian Sahara, including the desert around the oil-
fields at Fort Flatters 300 miles to the north and possibly the
land now occupied by the Great Eastern Erg itself. At the
point where we stood, close to the mountains, the brief torrential
rains falling amid the peaks at infrequent intervals descend so
fast and violently that the development of a mixed mosaic of
vegetation of the kind preferred by the locust for breeding is all
but impossible. Further north, where the wadi escapes from the
enclosing walls of the tassili on one side and the barrier of the
7,700-foot range of the Tedefast on the other, the desert is more
open, the rate and volume of the spate grow less and such moisture
as does not evaporate or seep into aquifers may provide the habitat
the Desert Locust likes. There have been heavy infestations at Fort

Flatters in the past, and with the outbreak of the next plague, twenty months later, there would be others.

Meanwhile we entertained the awe-inspiring thought that rain which fell in this spot when Christ was born is now being pumped up for the oilmen's Scotch and that but for the change of climate we should now be worrying not about the next waterhole but whether or not the wadi contained crocodiles. These are rumoured to exist still in the high combes of the Hoggars where water may persist from one rainfall to the next; but Pasquier said the evidence was scanty.

Past the last of the Irgharghar tributaries the black boulders gave way to rose-hued sand leading to a beautiful symmetrical bowl among the tassili's rocky foothills. Now the track was alternately grey and white as it first crossed shaly gravel and then sand, following a long gash upward through a succession of cols into yet more bowls and long shallow valleys. Two hours from Ideles it brought us to the edge of a sandy platform, a mile or two long, and wide, which had underlain an ancient plateau.

I was driving, as usual, with Abdel. Ours was the lead vehicle. We halted soon after mid-day at the foot of a bank of black scree to make sure that the rest of the party was still in touch. A plume of dust appearing round the bend of a mountain at the lip of the last col signalled the arrival of Landrover No. 2. It was driven by Fuad and contained Pasquier and Bennounou. It pulled up beside ours. With the vague idea of sighting Mahomet and Hadj I climbed a hundred feet up the scree and along the far side of a rock to a small peak. The change of height brought into view a desolate assembly of shattered rocks and hillocks of all lengths and shapes, most of them only a little lower than the one where I stood. Some were just heaps of rubble, brick-red and black as though stained by some transcendental fire. Here and there conical heaps of scree, like slag, represented a sub-stage of erosion, while amid them in the mid-distance north-eastward, at a higher level, lay a length of crag broken off at each end like a fallen super-cyclopean lintel. Its flat top, lining up with the tops of other isolated 'table mountains', indicated that all were once part of the same plateau.

As far as the eye could see, except westward along the track by which we had come, and to the south-east, where the Jebbel Telehrteba, last outposts of the Hoggars, lifted their lonely peaks, there stretched this same landscape of collapse and devastation, such as might have been left by a demonic ironmaster. I found myself thinking of Tolkien's *Lord of the Rings*. This was Sauron's country. But Sauron had his sky-riders. This had nothing. No bird or other living creature. No sound, except the diminishing whine of engines. No movement, except a pair of dust clouds feathering up behind two departing Landrovers. A mile away they turned the bend of the mountain and I had Sauron's country all to myself.

They returned an hour later with the missing one, all going very slowly. An embarrassed Abdel explained that, not seeing me around when he started back he had supposed me to have jumped in with Pasquier, as I sometimes did. Mahomet's car had a fractured spring. We lightened the load by shifting as much gear as possible to the other vehicles and limped on eastward over end-less gravel toward low lines of distant mountains which were already beginning to darken. In dusk we ploughed and lurched through a mass of the fine soft sand called fechfech, wheels alter-nately biting and spinning as we bore toward a long line of high sand dunes, the Erg D'Admer. It is about a hundred miles long and six to ten wide and the map indicates a track passing through it more or less directly to Djanet, 160 miles as the bustard flies. The temptation to push on toward the erg, camp late and get through it next morning in time to lunch and bath at Djanet was strong: but Bennounou would have none of it. He had 'smelled' timber again. Two old acacias supplied fuel for a fire. Soon the kettle was boiling, the radiance of the flames dancing over cheerful faces, the crouched figure of our guide bent over the cups, the jokes flying. Pasquier had been a little angry on hearing that there were no reserves of spare parts at Djanet, but soon recovered his good humour and, sitting on a jerrycan, commenced his nightly 'lecture'. It was to be our last night in camp. No recriminations were to be allowed to spoil it. So we talked the hours away, hands fanned out to the fire, then, when the embers burned low, turning ourselves about to feel the last comforting glow. It was good to be all together under the arc of the stars in this lonely place, and I had the first intimation here of the sorrow I would always have when a desert journey ended.

We were at breakfast the next morning when we heard the rising and falling sound of a motor in low gear wallowing through the sands from the direction of the track we had left the previous night. A few minutes later a Volkswagen minibus appeared. Its driver slid the door back and descended. He was a chunky-looking man with a flaming red face and gold teeth flashing in an ear-to-ear smile. He spoke in French with a powerful Italian accent, shook hands all round and beamed on the company as though in a *caffè* in his native Milan.

Down from the wagon behind him stepped a vision superior to any I had imagined when beguiled by mirages: a tiny woman, chic in sweater and trousers. She extended a graceful hand with a frank open smile stunning to our Arabs, starved for so long of female company, then turned and beckoned to a still smaller figure toppling out of the car behind her. 'Shake hands!' she commanded. The minuscule boy obediently went round the circle. From the rear window of the Volkswagen a still smaller copy, who, I suppose, was hardly more than two years old, peered out on a scene of a kind with which he was obviously going to grow rapidly familiar.

Meantime his father, a construction engineer, was talking to Abdel, who interpreted him to Bennounou. Bennounou told him that he should not be travelling alone. He said he always travelled alone. Producing a tattered and out-of-date Michelin map of the Sahara he pointed his finger at the dotted line of the track through the erg. 'Not good,' said Bennounou. 'O.K.,' said Goldteeth, 'See you in Djanet . . .' It was clear to me that he had misunderstood what Bennounou had said. It was also clear that like Bennounou he trusted his own judgment better than anyone else's. They all climbed back. Goldteeth gleamed through the window in a mighty, confident grin. In went the clutch. Off went the Volkswagen, its engine singing treble and bass as it lurched away in the direction of the erg. '*C'est un type, ça!*' said Pasquier. Bennounou, looking troubled, led us away northward, so that the dunes were on our right. There were no mountains now, only their stumps in various stages of degradation, some so advanced that all that remained were heaps of boulders. In times far distant the gravel will become sand, the wind will blow it against these rocks or any other obstruction – even the smallest, most withered of plants – and there it will pile up until new dunes form. Once we saw a 'wind-clock' composed of concentric circles traced in sand by the 'hand' of a drooping stem of a small shrub. As the wind caught it the hand turned around and around endlessly.

Toward late afternoon the reason for Bennounou's big swing northward became apparent. It brought us to a gap between the dunes and the tremendous crags of the Tassili n'Ajjer, bending from west to south-east in a bow 240 miles long. A new road, made up along the line of the caravan trail from Fort Gardel, led us down through black, rock-strewn ravines and over orange sand, below torn and jagged cliffs shaped into Gothic façades and machicolated towers and out to a broad valley filled with groves of krinkas, oddly ornamental. Their large glossy leaves signalled the approach of habitation. Soon we were among it: first a clump of grass huts at the edge of a dry river bed; then a honeycomb of stone houses clamped to the spur of a mountainside; then Djanet itself, consisting of houses and cavelike shops prettily edging a quasi-formal garden among groves of palms. We found the government guesthouse in the middle of it. Of zeribah style, its bedrooms were grouped round a tinkling fountain. No one had news of the Italians. Then we went up the hill to a roughish little restaurant and bar tended by a wall-eyed man and a tough-looking black boy about fifteen years old with a cigar stuck in the corner of his mouth. He had all the looks of a future big-time boss, provided, that is, he ever escaped from Djanet.

Goldteeth and his family arrived six hours after us. He said they had driven along the old track into the erg only to find that the dunes had advanced right over it. This had involved them in a detour of 140 kilometres, following the erg along its west side, as we had done. Getting out from the sand, he conceded cheerfully, had been pretty difficult. Behind his imperturbable, massive shoulders the signora looked as pretty and chic and cool as though she were just off to the Galleria. In cots in the bedroom the children slept soundly. I rather hoped for a good lie-in too. In vain. At 5 a.m. the voice of Abdel cried peremptorily: '*Allez!*

Allez! Allez! On va chercher les sauterelles.' Pasquier, who had slept among the krinkas, was waiting at the gateway, looking at his watch. We had an hour's drive over a sea of soft sand to where the locusts were awaiting us.

Wadi In Debirene lies about twenty kilometres south of Djanet and a hundred kilometres west of a lonely corner of the Libyan Fezzan. It is approached over drifts of glowing sand through which the short-wheel-base car made better progress than the rest of us in spite of its broken spring. A line of low cliff-like hills edges one side of the wadi. A distant fringe of trees marks the main course of the torrents. The rest is floodland. Otherwise, there is little to distinguish it from the rest of the plain.

Pasquier had visited In Debirene at the beginning of January, about three weeks earlier, and had noted as many as twenty-five hoppers and fledglings in single tussocks of bettina over practically the whole of the wadi. This, as he had told us, is a favourite plant of the Desert Locust and, being found at intervals right through the Sahara to the northern steppes, is a powerful aid to the insect both when swarming or, as now, marking time. A feature of In Debirene is that the tussocks are well but not excessively spaced. Little streamlets of sand run this way and that between the plants. Exposed to the sun, they are what seaside sand dunes are to the human sunbather – a wonderful temptation to bask. Many locusts were taking the opportunity to do so, crouched prostrate close to the plant, to which they could quickly retreat in case of disturbances or when ready to roost.

During Pasquier's previous visit the densest part of the locust population had been on the further side of the wadi, now it was on the other, suggesting a movement; but this, he thought, was more apparent than real and felt some form of destruction must have reduced the population on the former side. This would mean that the overall population was less, not because of migration for there was obviously plenty of food left. What then was the cause? The shadow of bird wings passing over a grey rock caused the Professor to look up. Half a dozen grey martins shot across the sky, then wheeled close above the bettina. From another stone Pasquier picked up the fresh remains of green hoppers – the legs and part of the cuticle – which had obviously been carried there by birds and partially eaten. These then were the marauders, an essential part of nature's check against the over-population which leads to gregarization and swarming.

During Popov's great survey he had observed repeatedly that, although the Desert Locust is a voracious vegetarian perfectly prepared to eat almost every plant that grows, it does in fact have certain preferences – bettina being one – and that when it finds them it will go for these rather than others. This narrowing of the feeding area can therefore be a first step toward concentrating the insect's population. A second occurs when the chosen vegetation is patchy, for this means the locusts tend to be gathered together in groups. Finally, when the drought is renewed and the favoured plants begin to dry out, the area of greenery correspondingly contracts and this also draws the hoppers together. Moving out of the dry vegetation into the green, they are thus brought into even closer contact. When all these conditions are found together, as they were in Wadi In Debirene, there is a battle between two opposing forces, and this is what we were witnessing. On the one hand nature seemed to be encouraging, indeed almost compelling, the insect to change its way of life from solitarious living to gregariousness and possible swarming; on the other hand she was using many means, including the birds, to keep the numbers under control.

The locust hoppers in the wadi were of practically all instars. Pasquier pointed out a number of nymphs bearing the dark markings of the early stages of gregarization; there were also hoppers in their fifth instar which were pure green and therefore solitarious. As the sun strengthened more came out to bask. The Prefect of Djanet, who had come out with us, proved adept at catching them with his bare hands or his hat. Abdel showed the correct way to do it by creeping up on a basking female, who in spite of her apparent somnolence was evidently using her immense angle of vision to good advantage. Twice she moved before, with gentle, stealthy 'Indian' movements, keeping his shadow always behind him, he was able to raise his net slowly and then bring it swiftly and successfully down.

Many others joined her in the killing bottle, to be sent to Pasquier's laboratory in Algiers for more leisurely examination.

Yet it was evident that here, as elsewhere in the journey, it was nature, not man, which had moved most effectively against the locust. The biggest number of nymphs, hoppers and adults that we found in any one clump of bettina was six. Nevertheless, Pasquier considered that although the locust population in Wadi In Debirene warranted no fears of an outbreak it was still sufficiently interesting to be worth a longer watch.

'You will have to remain a week,' he said, beaming quizzically at Mahomet, 'while waiting for a new spring for your Landrover to be sent down by air from Algeria. This will be an excellent opportunity for you to study the Desert Locust's habits.' Abdel was deputed to stay with him. They appeared to take the prospect cheerfully. Much of it, I fancied, would be spent in Djanet. Bennounou and Fuad were to make their way northward to Amguid and the middle reaches of Wadi Irgharghar and its branches.

We said goodbye to them next day on an airfield attended by a man, a boy and a dog. Two days later we were in Algiers and it was time to appraise our findings. We had driven across 1,200 miles of total desert and seen locusts in three places. Those in Wadi In Debirene were apparently on the point of dying out and indeed no more was heard of them. The two or three individual locusts we had flushed on the first day south from Silet were near to an important breeding zone in the southern Sahara and as the surveying team had seen more on their previous visit it looked as though there might have been a small migration, although here, too, natural predators, like Pasquier's palm lizard, might have accounted for the reduction. By far the most significant were the considerable numbers of locusts in Wadi In Attenkarer lying across the border between Algeria and Mali and quite close (as Saharan distances go) to the rocky massifs of the Adrar des Iforhas and Aïr, both of which send their floods far into the Tamesna, where there has been much breeding in the past. Nevertheless, all the specimens we had caught there had been solitaries and if Hadj Benzaza's count was correct there was no need, at that time, to assume that this population had reached a danger point. The schouwia, however, was very dense and because we had been working at the extreme range of our transport we had not been able to search it properly. It may have concealed more than was apparent. Moreover, there are other wadis of a similar character wandering across that part of the country and any of them could also have been harbouring just as many locusts as there were at In Attenkarer. The question was not whether any of these particular populations could multiply to the numbers needed for the first stage of a plague, for unless Benzaza had made a serious under-count this was unlikely. Natural losses would cut them down, leaving perhaps only ten per cent to migrate, or, as Pasquier preferred to say, displace. (Migration, he pointed out, implies some sort of purposive or instinctive action of a kind which is not, as we shall see, within the locust's capabilities.)

Where, then, would they go? Being solitaries, it was probable that their movements would not take them very great distances. Some might fly at night across the Tamesna to the Sahel country immediately south of the Sahara, or, depending on the winds, they could eventually land up in Mauretania to the west or in the wild lands around the Tibesti mountains in the east. Their progeny, or the progeny of their progeny down to a third or fourth generation, might well finish up in much the same area as those we had found. And this, indeed, is very likely what happened, for when the plague came again, in under two years, the Tamesna was one of the places where it started. In this sense 'our' locusts could have been the progenitors of the plague locusts. This led Pasquier to discuss the two theories about the prime causes of locust plagues which have produced some mild asperity among British experts. The first gives the influence of vegetation the most importance. The second holds that one plague is carried over to another by insects which have remained in the gregarious phase. According to those who advance the first theory the Desert Locust reverts to an essentially solitarious phase throughout its main out-break areas during a recession. Whether it again builds up into a plague then depends upon a complex of natural factors to which weather conditions, vegetation and the physical propinquity of one locust to another all contribute. The second theory puts its emphasis on meteorology. It supposes that somewhere or other in these vast areas one or more swarms are always on the move and that it is from these that, given favourable winds and weather, larger swarms are able to build up and diversify until plague again breaks out.

'In reality,' said Pasquier, 'these theories are not so contradictory as they look. It may be possible for a swarm to persist through a lull, continuously regenerating itself, although of this we have no positive evidence. And we know that solitarious locusts, having gone through the phase change, can eventually build up into swarms. It may well be a combination of the two events that precipitates a plague, but only – remember – *only* under certain conditions. So in making surveys we are not simply trying to find locusts. We are looking for evidence which will explain their behaviour to us and so enable us to find the best way to control them.'

Chapter 8

Wind, Weather and the Locust

Moses stretched forth his rod over the land of Egypt, and the
Lord brought an east wind upon that land all that day and all
that night; and when it was morning the east wind had brought
the locusts. And the locusts came up over the whole land of
Egypt, and settled on the whole country of Egypt, such a dense
swarm of locusts as had never been before or ever shall be again.
For they covered the face of the whole land, so that the land was
darkened, and they ate all the plants of the land and the fruit
of all the trees which the hail had left; not a green thing
remained, neither tree nor plant of the field, through all the
land of Egypt. *Exodus*.

Throughout 1966 and most of 1967 the Desert Locust remained,
as far as farmers were concerned, completely quiescent – a shy,
mild and not unattractive creature rarely seen by travellers even
in its desert fastness. An experienced team of Iranians and
Pakistanis searched from Teheran to the shores of the Indian
Ocean and back without finding a single specimen. In Saudi
Arabia we were not much more successful. About a dozen solitary
night-fliers were caught in light-traps set up on a high plain east
of Mecca and the Asir mountains and perhaps as many again
thumped into our lighted tent walls. Reports from Afghanistan,
Pakistan, northern India, eastern Africa and Sudan – all in
potential high-frequency breeding zones – were equally negative.

The one big informational gap was in the Sultanate of Muscat and Oman at the extremity of the Arabian peninsula. It was there that the previous plague had started following a cyclone and torrential rains; but the spring and summer of 1966 went peaceably with no untoward weather reports which might augur an outbreak. There was time, therefore, to consider some of the past research in field and laboratory on which future strategy would be based.

The suspicion that locust invasions are connected with wind and weather has probably always been deep-rooted in the minds of those who have suffered from them. In one of the versions of the Koran locusts are described, in a phrase both poetic and precise, as 'the teeth of the wind'. But the wind, to the peasant cultivator, is a local, not a geographic, phenomenon. He neither knows nor cares that it may have blown to him across a thousand miles or more. All winds and whatever they might bring – rains, beneficial or torrential, droughts, epidemics or plagues of locusts – are to him acts of God, or of the gods and the prophets who control and foresee all things. He will certainly, however, have noted their general periodicity and the directions in which they may be expected to blow according to the season. Father, son and grandson inherit and pass on a vague cumulative knowledge expressed, perhaps, in an equally religious belief, so that the wind blowing eastward from the Red Sea coincides with a locust migration some call 'The Army of Allah', while that blowing west from Rajasthan in northern India brings, for others, 'The Army of Rama'. Or religion may give way to magic, and especially the magical number seven, associated with good or bad luck, so that Saharan oasis dwellers, for example, assert, when questioned, that the Desert Locust plagues them for seven years running, then gives them seven years in peace. In such a belief we see the decay of primitive knowledge, based on observation, into a superstition still reflecting some degree of experience but in so distorted a way as to be valueless. All we are left with is the implicit fear, born of experience, of a vast insect cloud materializing out of the blue; that it comes with a certain periodicity; is probably to be associated with a weather change and is accompanied, more often than not, by a host of other ills which have plagued poor men from the beginning of time. It has been left to the scientist to find the way back to the

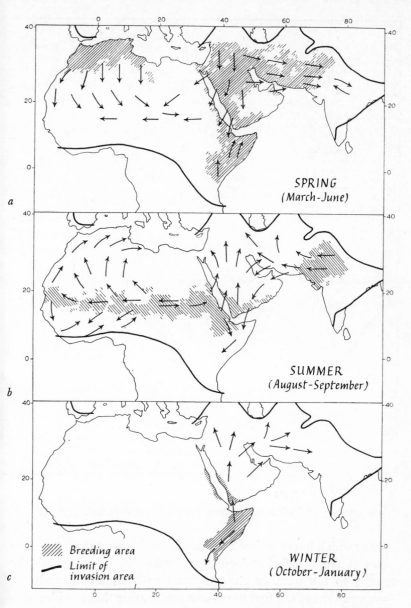

SPRING
(March-June)

a

SUMMER
(August-September)

b

///// Breeding area
—— Limit of
invasion area

WINTER
(October-January)

c

3 *a*, *b* and *c* Seasonal movements and breeding areas of the desert locust.

causes which seem to have been written so plain in *The Book of Exodus*, and this has been a long hard journey with many seemingly blind, if interesting, by-paths.

H

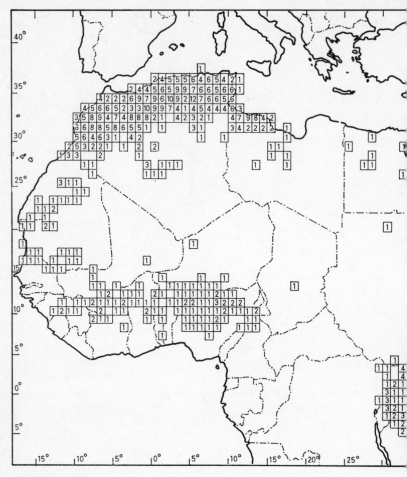

4 Frequency of swarms:

  a Frequency of April swarms over a twenty-five year period, 1939-63.

One which has exercised an unusual fascination over very many years is the belief that locust outbreaks can be systematically related to the weather changes apparently attending cycles of sunspots. One of the first to present some rather shaky evidence to this effect, in 1883, was the English entomologist A. H. Swinton. Some half a century later the grand old man of Indian locust research, Ramchandra Rao, took the theory a step further by offering a tentative forecast, after careful comparison of sunspot and plague years, that North-West India would for several years

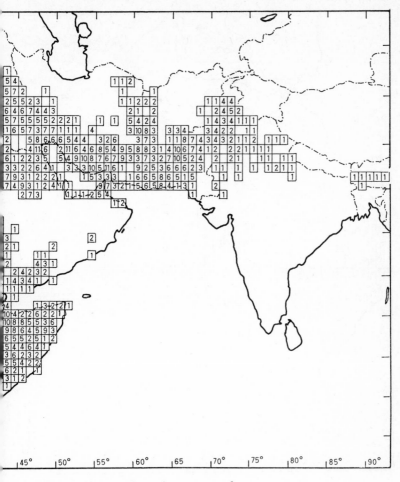

Boxed figures show numbers of swarms per degree square.

enjoy freedom from the plague. His hopes, alas, were rebutted by an invasion in the following year. In Algiers in 1942 my Saharan comrade Roger Pasquier, taking the previous century for study, also reached the conclusion that there was a definite correspondence between solar cycles and locust cycles. The North African invasions, he noted, were nearly always later, by up to five years, than those remarked by Rao in India, thus supporting his belief that outbreaks starting in the east extended sooner or later to the west and that swarms turning up in Algeria could well be of

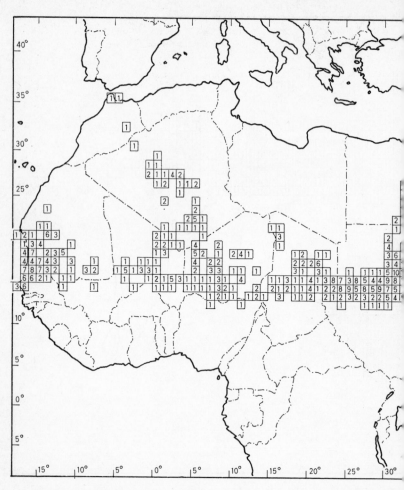

4 *b* Frequency of September swarms over the same period, showing them heavily concentrated along the line of the Inter-Tropical Convergence Zone,

oriental origin. Unfortunately, in the *mêlée* of war, his further conclusion, set forth in a subsequent paper, that the mainspring of a plague was basically a matter of ecology, reached only limited local circles. The point he was making was that, whether or not sunspots activated rainspells following droughts, the consequence of *any* climatic upset was bound to affect locusts. As the areas of desert vegetation expanded under the influence of rain, so would the Desert Locust's living space. As they withered and shrank, so would the habitable patches, forcing the insects into each other's

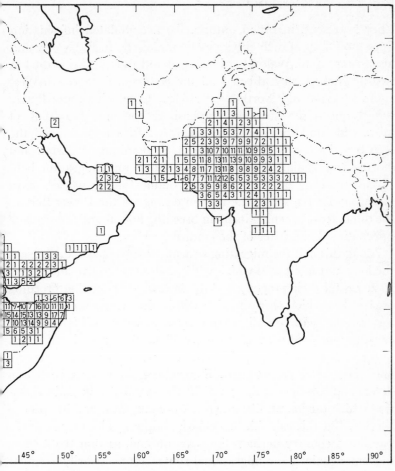

where the winds have gathered them. (Compare map 5b, p. 106.)

company. The one condition would pave the way for their increase and the other would control their density. These ideas, remarkable in their foresight, had to be rediscovered, so to speak, very much later and by then were anchored to meteorology by a new theory which had its own hard way to make before gaining general acceptance. This concerned the insect's migrations.

On page 99 the maps of the Desert Locust's principal breeding areas are marked with arrows showing the direction of swarm movements over a period of more than a quarter of a century.

They represent, in broad outline, the accumulated information from thousands of such entries made mainly by B. P. Uvarov and his assistant Zena Waloff between 1938 and 1963, covering the last two plagues as they affected all the countries from North-West India to West and North-West Africa. One point immediately stands out – the strongly emphasized corridor, hundreds of miles wide and thousands long, across Africa and, beyond the Red Sea, its continuation into Iran, Afghanistan, Pakistan and North-West India. This summer breeding and migration belt, resting squarely on the southerly confines of the world's most daunting deserts, shows the immense range of the Desert Locust, the difficulty of reaching its main breeding haunts and the size and international character of the control problem.

Along this belt the migrating swarms, maturing as they fly, seek the food and moist soils which will enable them to live and breed. If successful their progeny will fly on toward the spring breeding areas, from which *their* offspring will in due course move back to the summer belt. Not all swarms, by any means, will follow these broadly indicated routes, but this will be the general consensus.

It is easy to see how, by this seasonal merry-go-round, a plague, once begun, is perpetuated. Somewhere, in this vast area of desert and sub-desert, there will be the seasonal rain and vegetation which the locust, like any other nomad, must find in order to survive. But this rain falls irregularly, skipping a locality or whole parts of a country perhaps for years on end, so that the locust's movements along the migrational routes are necessarily irregular too. It is for this reason that nature has perfected a survival mechanism of extraordinary flexibility. As the rains die out, wherever it has been bred, the young locust takes wing, either in swarms or diffuse formation, or solitary, and, using a flying capacity unrivalled in the insect world, begins the long journey which will carry it to the next breeding area.

Until after the war it was thought that in making its big migrations the locust utilized its own sense of direction, which was presumed to bring it into certain specific breeding areas like that of its neighbour to the south, the African Migratory Locust, whose outbreaks had been tracked down to their source in the great bend of the Niger south of Timbuktu. At one time the desert in the bowl of mountains encircling the gloomy salt lake and marshes of the Jaz Murian in south Persia was suspected as a permanent breeding zone, but a search found nothing there. The only useful data about the migrations remained historical. The patiently compiled maps showed the routes taken by the mass of the swarms in the past and made it reasonable to assume that this is where they would be seen again. Unfortunately, the passage of many such swarms went unreported, so that again and again the trail of disaster followed with no chance to anticipate it.

One of the great difficulties in estimating either the size or direction of movement of a swarm is that, looking up into it, one sees locusts moving in all directions, thicker in one place, thinner in another, some settling, some taking off, everything on the stir, and often on so immense a scale that it is impossible even to guess what its limits are. A medium swarm may be several miles round when settled and when it takes wing the local configuration of the surrounding landscape may so channel the ground winds that the direction in which it finally flies may be quite different from that supposed by the local observer. It had long been known that wind affects the swarms on a local scale, as it does indeed many other insects. Some, like the common housefly, are so inhibited by it that they will avoid, if possible, taking to the air, while others, having done so, are hurled over great distances. The spread of the cotton boll-weevil, for example, is said to have been caused by cyclones and hurricanes; the fragile-seeming aphid has been whisked aloft and transported in great numbers for sixteen miles from the mainland to the isle of Memmel; the spruce aphid has been known to make a landfall in Spitzbergen, 800 miles by sea from its point of origin; and dragonflies, picked up and impelled onward by a severe hurricane in the region of Sumatra, have landed in the Cocos-Keeling Islands, 700 miles distant. Such flights are hardly migrations in the normal sense of the word, for most have depended on abnormal climatic conditions. On the other hand there is the gypsy moth, equipped with aerostatic hairs on which it floats and travels on the slightest breezes, thus ensuring its general spread.

The Desert Locust, with only its wings to keep it aloft and the winds to aid it, has far excelled any of those performances.

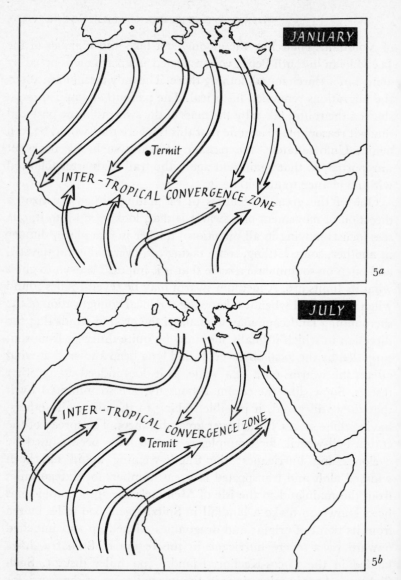

5 The Inter-Tropical Convergence Zone over Africa. (a) Winter position; (b) Summer position. Schematic surface windfields over northern Africa, showing the zone – the ITCZ – where opposing air masses meet, rise and may give off rain. Resulting vegetation enables the Desert Locust to breed and, under favourable conditions, multiply.

Over land a swarm may traverse as much as 3,000 miles, crossing perhaps the whole of Africa. Its flights over the sea are even more impressive. A map prepared at the Anti-Locust Research Centre shows them scattered thickly over the Persian Gulf and the Red Sea, and along the coasts between, with a further considerable group of sightings off the West Coast of Africa. Ships' reports have occasionally provided the first warnings of invasions of eastern Africa by swarms coming in over the western Arabian Sea; but it is in the Atlantic that the greatest flights have been recorded. Some of these have been so long as to seem incredible. Thus there have been three reports of Desert Locusts landing in numbers on ships in mid-Atlantic at distances of 1,400 to 1,500 miles from the West African shore, and flights to the Canary Isles have proved to be well within their range.

The longest of all flights seems to have been one from the Canaries to the Scilly Isles and the west coast of Ireland. This would appear to have involved the insects (which landed alive) in being continuously in the air for about sixty hours, since there is no land in between. Theoretically, even allowing for a strong tail-wind and a good deal of gliding, such a feat is impossible, at any rate by any known standards. In the laboratory wind-tunnel the longest recorded spell of continuous wing-flapping is about seventeen hours and individuals in even the fattest batch of swarming locusts so far captured in the field are unlikely to have had a fuel-reserve capable of keeping them in the air for more than twenty hours. Perplexed scientists have therefore fallen back on an old belief that as the swarm reaches the end of its tether and descends wearily to the sea, the first to alight drown and form a floating island on which the others are able to rest – and not only rest but feed, picking off weakling fellow-survivors so that they can restore their own fat reserves for the next stage of the flight. There is only one record, and that a hearsay one, of such an event ever having been witnessed – by the crew of a ship in passage between India and Tenerife in 1649. Nevertheless, floating masses of dead locusts have been reported far out in the Atlantic as recently as 1954; and no one has provided an alternative explanation.

Although it was thus evident that the wind could blow swarms far off-shore, most of the older records could have been regarded as describing isolated incidents with no particular bearing on the routes they might take over land. In fact, there is no visually obvious resemblance between the locusts' land movements and those over the sea, for whereas the latter are bound to be more or less continuous the others are broken into daily stages. In eight to ten hours daily flying they may move any distance between, say, ten and one hundred miles. Their long-range movements are therefore spread over weeks rather than days. Until just before the outbreak of the 1950 plague no one had realized that meteorology could provide the common factor and explain why these locusts were to be found where they were. Nor was this surprising, for even now in the great deserts meteorological stations are very few and far between. There was fortunately one important exception. Around the Gulf of Aden, a number of stations had been set up during the war in order to aid the Allied air forces in submarine control. A glance at the maps of the locusts' breeding areas, particularly those of the summer seasons, will show that all the country around the bottom of the Red Sea and on either side of the Gulf is a cross-roads of the migration routes. So often has the despairing farmer seen locusts apparently flying in from the sea that many Arabians believe that they originate in the stomachs of whales. I have heard a villager ask a Saudi locust officer if they were originally a kind of fish; and in Persia, where they are seen in flight across the Persian Gulf, the Gulf of Oman and the connecting Strait of Hormoz, their common name is Sea Locusts. For control workers in eastern Africa the picture was complicated and partly obscured by the complexity of the local winds, which might blow a swarm off-shore and bring it in again elsewhere, to be mistaken for a new swarm. Yet it was in this area that the mystery of the migrations was cracked and a new approach to the whole locust problem born.

Once more it meant that men had to go back to a study of the winds, taking not only those of a locality or even a country but of the whole known range of the Desert Locust's operations between the Himalayas and the Atlantic. The idea that the directions of these great insect swarm movements might be connected with weather had been dabbled with before, notably by an intrepid French entomologist-explorer, Volkonsky, who with his wife,

before the war, made a number of long-distance journeys by
camel through the wastes of the Sahara, tracing the locusts'
wanderings westward after summer-autumn breeding in the wadis
of the bleak gravel tropical plains south of the Hoggar Mountains.
Whether he could at this time have obtained enough evidence to
elevate a tentative idea into a workable hypothesis is doubtful.
He died, in any case, a victim of his own beloved desert, of an
illness exacerbated by the Sahara's extremes of day and night
temperatures before he could attempt to do so.

Meantime the study of meteorology, quite apart from that of
locusts, had made a long step forward with the development of the
theory, originated about the time of the First World War, that the
climate and also the terrain of a considerable part of the world is
to be explained by the movements of the winds in what is now
usually referred to as the Inter-Tropical Convergence Zone.*
Basically, and in very general and unscientific terms, it is a
climatic band of varying width encircling the globe in the approximate region of the Equator. It does not remain in a fixed position
but moves seasonally northward and southward, following the
sun. Within its somewhat indeterminate boundaries two opposing
masses of tropical air meet, rise and form clouds giving off rain,
which is at first light and erratic and then often heavier. The
winds on the north side are known to the sailor as the North-East
Trades; those on the south are the South-East Trades. Obviously
where they converge they will be blowing generally toward the
west; conversely, at times when they gain a westerly component,
they will be directed toward the east. This happens regularly
when the ITCZ makes its winter southerly movement across the
Equator. The North-East Trade then becomes a North-West
monsoon and the South-East Trade becomes a South-West monsoon. The degree and intensity of wind convergence are not the
same everywhere, for the section of the Zone with which we are
concerned, stretching right across Africa and North-West Asia, is,
after all, some 7,500 miles long, giving plenty of room for local and

* It can also be called the Equatorial Trough or the Inter-Tropical Front
or the Inter-Tropical Discontinuity, each of which in its own way describes its
features. I will here call it the ITCZ. Later, when we meet it over Ethiopia, I
shall just refer to it as 'the front'.

large-scale disturbances which may upset the general pattern. But in some parts the seasonal changes are very marked indeed and have a profound effect on human life. Among Swahilis, for example, the wind at the change is known as '*tanganbili*' – 'two sails' – meaning that the Arab trader who has been bowling along the coast with the aid of a following wind in one direction can now go turnabout and enjoy a favourable wind to carry him home. That these changes of wind also frequently coincided with the arrival and departure of locust swarms was a matter of local knowledge empirically and painfully acquired by many generations of farmers. Synoptic meteorology – the study of day-to-day atmospheric movements – has not only confirmed the correctness of their observations. It has established a direct link between the movements of winds and the movements of locusts throughout the eleven million square miles of their breeding and invasion areas.

The introduction of detailed meteorological studies into locust research was partly due, appropriately enough, to a young English scientist's enthusiasm for flying and gliding. R. C. Rainey's theory, now amply proved, is that the major movements of all Desert Locust swarms take place downwind into zones of convergent surface air-flow where rain is likely. Put more romantically, as befits this fantastically adaptable creature, it means that they ride the winds where the winds list and in doing so may find what they need for their self-perpetuation.

Rainey had come into the post-war anti-locust campaigns in an indirect way and with a fresh mind. Using experience gained very largely in other fields, he was able, in his own words, 'to put together the bits of information no one person ordinarily gets to know about'. At Imperial College, where he took his degree in biology, he had been an enthusiastic member of the college gliding club, 'picking up the meteorological aspect', as he has said, 'more or less at bar-talk level'. His first job – with the Empire Cotton Growing Corporation in the Transvaal – brought him into entomology as a student of cotton stainers – flying and migrating pests whose activity in the cotton could cause serious spoilage. As a flier, he found himself interested in their flight characteristics, only to have his observations stopped by the outbreak of war. The indirect outcome of this was his return to meteorology. After being told that his eyesight was not good enough to allow him to be a Service pilot, he was drafted into the meteorological branch of the South African Air Force and spent his war in East Africa, Somaliland and the Middle East.

There Rainey had his first experience of the Desert Locust, and, although unable to take an active part in the immense control campaigns then being waged as part of the war effort, it was at this time that he began to think of the possible connection between the locusts' movements and those of the prevalent winds. Chance put him back on the subject when he had returned to the Cotton Corporation after the war.

'I had written,' he recalls, 'a paper for the Royal Meteorological Society on the meteorological aspects of gliding. John Kennedy, a famous locust authority, who had read it, met me one day at the Entomological Society in London and asked me to meet some locust people. They were Dr B. P. Uvarov, Director of the Anti-Locust Research Centre, Dr D. L. Gunn, his deputy, and Miss Zena Waloff, whose work, with others, on locust migrations had been so important in the wartime campaigns. Gunn said they were looking for the loan of someone who would be a sort of hybrid between an entomologist and a meteorologist, and asked me if I had done work intermediate between synoptic meteorology and micro-meteorology. I had, in the shape of the Cotton Corporation job, and they invited me as a first step to go off to East Africa with Miss Waloff to see what relationships could be worked out between swarm movements and the weather in general.' This was in 1947, a year significant in locust history in that it marked the beginning of a brief break between the plague which had broken out in 1940 and the appalling upsurge of a new one which was to bring ruin to thousands throughout the next decade.

This was indirectly fortunate, for with the swarms beginning to die out Rainey was able to begin rooting around in the records of the Meteorological Office at Aden for weather data to compare with Miss Waloff's observations on the movements of the Desert Locust around the Gulf. These records were exceptionally detailed and by co-relating them with Miss Waloff's records he was able, in his own words, 'to begin to make some sense' of swarm movements. 'We found that in October of 1947, for example, a swarm had come right across the Gulf at a time when the wind was in the right quarter to carry them; and in July another had evidently been blown offshore and back.' Other weather records fitted the picture the two scientists were now able to construct. Having realized that the direction taken by the locusts must be dominated by the wind, it seemed logically clear to Rainey that in moving downwind vast numbers must inevitably be channelled on to the moving belt of the Inter-Tropical Convergence Zone; in other words, that when the swarms set off from one breeding area to another this important part of their life cycle was not just an affair of chance or instinct, as supposed, but was geared directly to the winds. Within the Zone there are, of course, many complexities. Winds are going out of it as well as going in, so that a Desert Locust swarm may travel through many a zigzag and even, at periods, return almost to the point from which it started. Nevertheless there is a general displacement and this word, rather than migration, has come to be used for the long journeys on which the insect must embark when the time for movement has come. Whatever the word preferred, it became increasingly clear from comparisons between the known weather at certain times and Miss Waloff's swarm records for the same periods that there was a degree of co-relation which stretched far beyond coincidence.

Back in London, Rainey discussed his ideas with Uvarov; but the proof of the pudding, as both realized, could only be found in the deserts – where, however, there were now no swarms. For a year the majority of the reports reaching the Anti-Locust Research Centre spoke only of isolated breeding or, at most, of swarmlets which had apparently come to nothing. But although farmers from India to Morocco and Senegal might breathe more easily, Uvarov, as always, urged that this was the very time for intensifying field research. The outcome was the setting up of a field research service which he named the Desert Locust Survey, covering the huge area from Kenya to Saudi Arabia and the Persian Gulf.

The new survey team was soon to be involved in a tougher and longer battle with the locust than any of its predecessors. But of this there were no immediate signs. Before the storm broke Rainey was offered a job as senior entomologist with the survey at their then headquarters in Nairobi, where one of his first steps was to

arrange for a daily weather map provided by the local Meteoro-
ogical Department, showing the whole survey area and including
he daily positions of the ITCZ. He had not long to wait to start
comparing that with locust swarms, for the 1948-9 recession turned
out to be the shortest on record. By February 1950 Saudi Arabia,
Yemen, Aden and Sudan were again being invaded. By July
reports from these countries and from India, Pakistan, Iran and
Somaliland, when plotted on the maps, showed the distribution
of the locusts to be matched very closely with the position of the
ITCZ, then fairly stationary in its northern position.

On 26 September he received the first report of a swarm moving
south in Somaliland. It coincided precisely with the movement of
he ITCZ. This still left it necessary to gain the second vital half
of the 'proof' by translating the map plottings into a forecast. The
opportunity came within three weeks, when, in mid-October, the
Meteorological Office stated that the ITCZ was rapidly moving
outhward. No swarms had yet arrived in Kenya and Rainey
hereupon sent out warnings to the locust units and administra-
ion in the Northern Province that they might be expected at any
ime. He recalls the ensuing waiting period as one of the most
anxious in his professional life. It ended with a signal five days later
reporting that the locusts had arrived.

The single example of a successful forecast, however, was not
enough to prove a rule. In the battle with an insect of such far-
ranging powers the logging of all observed movements and the
environments in which they occur has had to be carried out in
many places over many years and it was not until the mid-1950s
hat a systematic twelve months' analysis, carried out conjointly
by the World Meteorological Organization and the Anti-Locust
Research Centre, put the truth of Rainey's theory linking the
migrations with the ITCZ beyond all further doubt. The answer
o one apparent set of exceptions to the downwind rule was made
clear to him one January day when the BOAC Hermes on which
he was flying out to East Africa from London was held up at Rome
airport. This delay, it was explained to the sceptical and annoyed
passengers, was due to exceptionally bad weather, with contrary
winds, over the Mediterranean. Rainey alone among the passengers heard the news with enthusiastic interest. At this time of the

year, when the ITCZ is in its southerly position, the polar front advances behind it and cold northerly winds blow far into North Africa. In theory locusts could not possibly migrate northward, even if they could fly against the wind, for the temperatures would be too low for normal flight activity; yet spring swarms originating in the south of the Sahara had consistently appeared in the North African coastal regions. Speculating on possible explanations, it had already occurred to Rainey that the Desert Locust, being a supreme opportunist, might have been able to take advantage of the air movements in just such a weather upset as was holding him up in Rome. This, he conjectured, was being caused by a deep depression as the result of which the northerly winds were being interrupted by a temporary spell of warm south-westerlies. This proved to be the case.

It was also the case, although he did not know it at the time, that a big northward movement of locusts across Africa had coincided with the same depression. Once the temperature had risen sufficiently to encourage the swarms to fly, the temporary south-westerly winds had done the rest. Such periodic depressions are normal during the Mediterranean spring and since they are commonly accompanied by rain they often provide precisely the conditions required for breeding. It is also known, or has at any rate been frequently noticed, that with the advent of rain the locust's sexual maturation is considerably speeded up. It is coincidences of events like these that help a plague to build up and keep going. In the early spring of 1955 a similar depression, deepening under the lee of the Atlas Mountains in south Morocco, caused a northward and eastward spread of swarms leading to practically the whole of the North African coastal farmlands almost to the Nile delta becoming infested. Utilizing such favourable spells, the Desert Locust can make short cuts enabling it to keep itself going on intermediate breeding circuits within sections of the whole vast invasion area. It is not necessary for migrant swarms to traverse the whole or even a great part of it. Weather and vegetation permitting, they can keep their numbers up to plague dimensions in one region while others are for the time being free. Then, given the right circumstances, they can re-infest the rest. (There may also, of course, be independent local outbreaks. The problem

9 Young adult gregarious locusts, although winged and ready for flight, may retain the habit of marching for several hours, even days, after fledging. They then take wing as a swarm.

10 Fledglings massed on wild palm. The infestation, in Ethiopia, covered several square miles and was successfully controlled.

then is to prevent them spreading and joining.) During all plagues this shift of intensity from region to region has occurred repeatedly. The importance of Rainey's discovery is that it has helped to narrow down the areas in which outbreaks are to be expected, and though this hardly seems much comfort when one recalls the huge size of the recession territory, there nevertheless remains the possibility of still further whittling down. Instead of looking for specific breeding areas, the Desert Locust expert, armed with improved meteorological knowledge and the statistical evidence of past outbreaks, now seeks for the high incidence areas and watches what occurs there. It is a remote and lonely watch, but without it the Desert Locust is unlikely ever to be completely defeated, though he may be contained.

Meantime, while the evidence in support of Rainey's theory accumulated, he and others were carrying out studies, made largely from aircraft, and also with the aid of ground photographs, of the actual structure of swarms and the behaviour of the locusts composing them. This is a highly complex subject, in which a lot of questions remain unanswered. It is important because it directly affects the question of control by air spraying. By being able to estimate, for example, the density of a swarm it is possible to give a dimension to the threat. But it is also desirable to know the forms in which the threat is likely to come, and these are manifold. They vary from swarms which may be only a few yards deep from the top to the bottom layers, yet perhaps cover several square miles, to others towering four or five thousand feet. A single swarm may take both forms and any intermediate one at various times in the day, and as the shape changes so of course does the density. Rainey has reckoned that a square kilometre of a well-packed swarm observed in East Africa has contained 40 million locusts. Estimates for others have ranged up to 150 millions per square kilometre. At such rates a large swarm covering, say, 100 square kilometres could contain over 10,000 millions, but not all will be flying at once. Often some will be settling while others will be taking off, so that the swarm, however uniform it looks at any one moment, is really progressing in a rolling motion, one part constantly replacing the other in the air as the whole body of the swarm moves forward.

I

A stratiform swarm may be so dense that it is possible to see at most a hundred yards into it from the edge. Looked at from the air, it may blot out the ground below it. Seen from above and at a distance, with the sunlight striking it at an angle, it has an intense and unmistakable glitter caused by reflections from the wings. Then it may be seen to billow. Parts rise up to become 'shining, gauzy veils or columns' – I quote Rainey – 'appearing, moving and disappearing with the changing angle of viewing. On another occasion it was noted that, with the passage of the observing aircraft, a point of intense specular reflection travelled up such a column, along perhaps a quarter of a turn of an apparently spiral path.' Sometimes the swarm had the appearance of a net of more or less regularly spaced wispy elements, often forked or Y-shaped. In both effects – net and veil – it conveyed the impression of something evanescent, ethereal. As it assumed them the swarm ceased to be stratiform and became cumuliform.

'Viewed from a lower altitude, against the sky, cumuliform swarms often resemble smoke, and from a distance have at times been mistaken for bush fires, but have usually been distinguishable by a characteristic fibrous or streaky texture, with groups of high-flying locusts sometimes giving the impression of clouds, streaks, or wisps relatively separate from the main form.' One observer likened it to the wispiness of mist freshly forming in a valley, or to the pencillings of tobacco smoke drifting in the still air of a closed room.

'On occasion an effect as bold as a charcoal stroke has been seen against the sky, sloping downward at a marked angle to the horizon, and subsequently fading within a few minutes as the aircraft approached the swarm. Another effect, lasting for several seconds and seen on two occasions while flying past large swarms at a distance of several kilometres and at a height below that of the topmost locusts, has resembled a thin plume of smoke from a moving train on a calm day.' Yet again, the shape of a cumuliform swarm has been compared with a number of great pillars of peculiar smoke, a mile apart, rising from the general line of the swarm. In Kenya, another swarm erupted into pillars which rose to a height of 1,500 feet. As some pillars disintegrated others formed, all moving together in the same direction as the main swarm, while from the valley floor rose dust 'devils', towering to great heights then falling away again.

Pillars of locusts and 'devils' of dust are both caused by vigorous convective up-currents of air. Occurring during the heat of the day, they have been known to lift scattered locusts to the level of a plane flying at nearly 10,000 feet or over one and three-quarter miles. One of the most remarkable things about a swarm of locusts, nevertheless, is its cohesion. However strange the shapes it may take, however many the columns and curtains which may appear and disappear, it seems to be governed above all by the need to stick together. So great are its numbers and so dense at times, that it is thought that a swarm may even create a measure of its own air turbulence by means of the heat it gives off as it flies.

Photographing from below with cameras adapted to give double exposures, so that the insects in view are shown in two consecutive positions close together, observers have detected another characteristic which on the face of it also seems disruptive. The enlargements from the negatives showed that instead of all flying in the same direction, as one would expect, the swarm included large numbers of individuals and groups headed all over the place. In a swarm which as a whole was travelling west-south-west, for example, a large proportion of the locusts composing it were flying toward nearly all points of the compass, some even going in a diametrically opposite direction to the general drift. These were in addition to those taking off and landing upwind – their invariable habit. Yet in spite of the apparent confusion as the groups within the swarm crossed tracks, there was no sign of breaking up. On the contrary, the edge of the swarm remained quite clear-cut, as group after group, having reached it, turned as though on some mysterious order and flew back into the main body. In the same way those which had reached the front would also wheel back, giving the impression of two contrary streams. Only at the rear edge, in general, would the courses of individual flights coincide with the general track of the swarm. In a stationary swarm the effect is that all the interior flights of all the locusts composing it virtually cancel each other out. Winging industriously at cross courses they end where they begin. Such a swarm is waiting for a wind, which may itself then box the compass and bring the swarm back days later to its starting point. If this goes on too long, over too arid a stretch of country, the swarm may die, just as it must do if the wind consistently carries it forward, as often it does, into country where its needs are absent. Yet so long as it lives, so long will it appear to strive for the cohesion on which the best chances of survival depend. And this cohesive drive will be shown not only in the swarm but among the hoppers to which it eventually gives rise.

Chapter 9

On the March

The word of the Lord that came to Joel, the son of Pethu'el:

> Hear this, you aged men,
>> give ear, all inhabitants of the land!
> Has such a thing happened in your days,
>> or in the days of your fathers?
> Tell your children of it,
>> and let your children tell their children,
>> and their children another generation.

> What the cutting locust left,
>> the swarming locust has eaten.
> What the swarming locust left,
>> the hopping locust has eaten,
> and what the hopping locust left,
>> the destroying locust has eaten.
> Awake, you drunkards, and weep;
>> and wail, all you drinkers of wine,
> because of the sweet wine,
>> for it is cut off from your mouth.

> For a nation has come up against my land,
>> powerful and without number;
> its teeth are lions' teeth,
>> and it has the fangs of a lioness.

> It has laid waste my vines,
>> and splintered my fig trees;
> it has stripped off their bark and thrown it down;
>> their branches are made white.

As it was in the days of Joel, so it has been throughout history. The very splendour of the language, in which he evokes a picture of a people waiting anguished while the air fills with the sound of the locusts' wings and the crunching of their jaws while sky and land darken in the shadow of their numbers, tends to obscure the fact that all fourteen verses of 'The Devastation by Locusts' contain a precise description of what happens during a plague. Here is the *exact* picture of what the advancing hordes of locusts and hoppers look like – seen under a magnifying glass even the comparisons with horses and lions become far from fanciful – and this is *exactly* how the hoppers behave when on the march.

There is the same accuracy in the *Proverbs'* observation that 'the locusts have no King, yet all of them march in rank'. These were the eye-witness observations of men who had seen the enemy not once but many times. The terror of the locust was bitten so deeply into all of them, prophets and people alike, that the reiteration of its name and its attributes – cutting, swarming, hopping and destroying – is held to be enough to induce penitence. In singling out the marchers – the hoppers, each marching unjostled 'in his own path' – Joel, however, is also noting one of the strangest aspects of the Desert Locust's life cycle with which scientists have had to grapple in seeking the best means of control.

The disposition to march is a feature, primarily, of the insect's gregarious phase. The verb, like the prophet's majestic verses, conjures up a picture of companies, regiments, whole armies on the move; and this, in a sense, is what they may well be as one small group of hoppers, a day or two old, links up with another, or a band, already fairly large, joins perhaps with another one larger yet. At its maximum stage of expansion such a band may be ten miles wide and many more long and will march as one imagines the Mongol hordes of old did – a great mass on the move from dawn to dusk or later but always with some groups or individuals within it coming to temporary halts, to feed, to rest, to bask in the sun while, to potter and peer, to fidget and fight, yet always to join the forward movement again.

One must not, however, stretch the military comparison too far. Though such hopper bands have been reckoned as responsible for some eight per cent of all the losses caused by Desert Locusts, these marches are not for plunder, nor are they, as once thought, set off by some kind of communal stress – starvation, for example. Studies in the laboratory have shown the *Schistocerca* hopper marching just as cheerfully on a full stomach as an empty one. Starvation might increase marching for a short period, but if continued for long, produces a reduction. Stress, in other words, apparently has nothing to do with it.

The locust hopper is impelled to march not by fear, nor by a blind conviction that there are better pastures over the hill, but by its own behaviour pattern. It marches best when the temperatures are also best for its living and breeding generally, and worst, covering the least ground, when it is either too hot or cold. The more its numbers, the more it marches. The bigger the crowds, the bigger the mutual stimulus – and this fact is of immense importance in studying the gregariousness of locusts in all their aspects, including their transition. Thus, a small band may march only a few hundred yards a day, or perhaps a mile in all in the course of its growth from hatchling to adult; whereas big ones have been known to move a dozen miles, at the rate of nearly a mile a day, in the same period. It is almost as though, in some mysterious locust fashion, they had cheered themselves on.

Basically and evolutionarily, the march is connected with feeding in that, once in a mass, the insect has no choice. In a given feeding ground, containing a large number of hoppers, each, if they are scattered around, will be able to find what it wants in its own locality. It has only to browse to eat. In the desert wadis one finds many such creatures, from gazelles to dromedaries, which will only leave their pastures if disturbed. Gregarious Desert Locust hoppers are not in this situation. Living in dense groups, often with big areas of arid country between one group and another, they must move if they are to eat enough. What this locust has done is to evolve one jump ahead of some other species. Whereas the hoppers of the Moroccan Locust, for example, tend to eat themselves more or less entirely out of their food supplies before moving on to the next suitable vegetation, the Desert Locust hopper only rarely consumes everything. It eats as it marches and the march, in the general mass, is constant. Evidently, therefore, the marching arose out of necessity, but why should it be in such big bands?

As each band marches it tends to snowball by picking up others, and some observers believe that this too is another way in which the locust ensures its own survival as a species against attacks by predators. It is a complicated picture, for although on the one hand a big band seems likely to attract predators, this is offset by the fact that the majority of predators – particularly birds – have their own territories in the bush. If a locust band comes into their territory they will feed on it, but they will not necessarily go outside their territory in order to find the band. Many birds, moreover, seem afraid of a big band, particularly of the moving mass at the front where it is packed most strongly. They will eye it but tend on the whole to leave it alone, preferring to pick on the stragglers at the end. Small birds appear to be particularly nervous of the Desert Locust hopper when they meet it *en masse* and can even be scared by the jumps of hoppers in quite small groups. John A. Hudleston, a member of the British-supported Desert Locust Survey, who observed birds attacking a relatively small band in Ethiopia, concluded that in a large one of the kind which occurs when egg-laying has been heavy, the appetite of the attackers would have been sated before the band was destroyed. That such a band, through its sheer numbers, can swamp the capacity of the predators is also the conclusion of Dr Peggy Ellis, one of the leaders of the same field survey. She says: 'I think living together in a big group is a way of saturating the hunger of all the predators likely to encounter and endanger it. It seems likely that fewer locust hoppers are eaten when they are on the march in big bands than when they are scattered and can be attacked by a greater number of predators.'

Dr Ellis and Clifford Ashall, both of the Anti-Locust Research Centre, have described the strange desert scene in East Africa as it came to life at successive dawns during the plague year of 1953. 'Before first light the majority of hoppers were roosting. When colours first became easy to see (some twenty minutes before sunrise) a few hoppers began to walk down the plant stems or trunks, battling their way over other resting hoppers. As light intensity increased, so did the downward movement, until whole groups became involved. Only occasionally did a hopper fall.' Even in quite large roosting places, spread over hundreds of thousands of square yards, these preliminaries to the march began at the same time everywhere, except among some hoppers on the shady sides of leafy bushes which got going somewhat later. This suggested that it was the intensity of the light that triggered off the movement. If, for this or some other reason, the laggards were too many, however, those on the ground would clamber back into the bushes again, to wait, perhaps for forty minutes, when at last they would all stream down and, forming into dense columns, begin to move off, in roughly the same direction as they had taken on the previous day.

Once a good stream was in progress, the hoppers remaining in any bushes they passed soon joined in, all walking and packed so tightly that they almost touched each other. A rate of advance of twenty feet in five minutes, the observers noted, was not unusual and a band might shift as much as a hundred yards during this early morning period of marching, which rarely lasted for more than two hours. One thing very evident was that the bigger the band the faster and farther it marched. Another was that although individual groups might seem to be going all ways, sometimes even at right-angles to the head of the band, all as a whole would maintain the same direction, and do so for days at a time. It was as though the band progressed like an amoeba, dividing to flow round obstacles but forever coming together again. These erratic courses seemed to be taken chiefly by those in the middle. At the back the stragglers appeared to march on much straighter courses and gave the impression of trying to catch up with the others. Those at the front also seemed as though anxious about losing contact with the rest. If one found itself ahead too far it would stop and wait for the rest to catch up, or might even turn back to them. It looked not as much as though those at the front were leading as that they were being propelled onward by the massed impetus of the others close behind.

What governs the direction of a band is a question still not

finally settled. Dr Ellis and Clifford Ashall concluded that the
wind was an important factor. They noticed that the smaller
hopper bands (under twenty thousand) tended to move downwind
fairly consistently, and even the larger bands, although apparently
able to move on any chosen orientation, would still tend to fan out
on the downwind side. The distribution of the vegetation, thick or
thin, gappy or even, also seemed to affect their line of march considerably, particularly if there was a sudden break. Faced with an
open space, a massed band of hoppers seems to get an attack of
agoraphobia and, unless in very large groups, will start out over it
only with the greatest reluctance. This was very noticeable in the
East African bush country, where a luxuriant growth gives way
every now and again to an area where everything has dried out
and cover is virtually non-existent. 'This may not be very big,'
said Dr Ellis, 'only a few hundred yards perhaps. When the band
comes to it the leaders will start marching across and then suddenly
will all turn round and come back into the bush. On the other
hand, narrow areas, such as a road, may channel their movement
because they find, after a trial, that it's easier to walk along it than
to push on through the bush. So this would be an overlying
factor in deciding their direction.'

Once, on a newly graded road fringed by sandy ridges, the
observers watched while a band made repeated attempts to cross.
First, some hoppers would start out, then, finding the rest still lag-
ging reluctant in the bush, they would stop, turn round and come
back. As they walked on the road surface, they were prevented,
by the ridges, from seeing any of their fellows on either side
and this seemed further to deter them. Finally, a large group
crossed both ridges and road, only to find themselves alone
because the rest of the band had meantime decided it was time for
their morning pause. Fifteen minutes later 244 hoppers walked
back across the road.

This has provoked a fascinating question. Because of the sand
ridges these 244 could not possibly have *seen* they were not being
followed. How, then, did they know? Dr Ellis said: 'It is one of
the puzzles. You come across odd cases where you feel the locusts
are in communication with each other in a way we don't yet
understand. Very occasionally in the field we have found that

when two hopper bands get fairly near to each other their paths alter quite suddenly and they turn toward each other. If they are, say, forty or fifty yards apart at the time, they can't possibly have seen each other. It's possible they could smell each other, but we have never been able to show in the laboratory that locust smell is attractive to other locusts, although in the field it may be different. It is possible they hear each other. The human ear can certainly hear a hopper band moving in the bush – it's a very curious sort of pitter-patter, a most extraordinary effect when made by a big band. Even an individual locust moving around in a bush makes a very tiny noise that you get to be able to hear. So it could be sound, or a mixture of sound and smell. Communication certainly happens when there is no question of sight.' In the laboratory, Dr W. T. Haskell, Director of the Anti-Locust Research Centre, did some work with adult locusts on their possible range of hearing, but they couldn't apparently hear each other when more than a few feet apart.

Although rarely more than seventy per cent of the band are on the march at the same time, there are three main periods when most of them are stopped. The first usually comes within a couple of hours of dawn. This is a moment very noticeable to the human traveller. The sun has come up perhaps over a mountain ridge, the night's freshness has vanished but the heat is not yet intense, the long shadows of the dunes shorten and for a brief while one has a sense of the most wonderful well-being. It is at this time, when the sand has been coaxed into warmth, that the locust hoppers halt their early morning march. Some, with bodies turned sideways to the sun, begin to bask. Others, the majority, collect in groups so dense that the hoppers are piled two or three deep. The groups are constantly heaving with movement, surging up into volcanic jumping outbursts as those below leap up to exchange places with those above. One leap sets off a frenzy of leaping until the whole group is at it, only to subside into a general restless kicking, fidgeting and pottering. Although this is therefore not a resting period, it is evidently connected with the sun. A shadow cast over them may cause them to move, and grouping will only normally – and then not always – occur in shade if there is a strong wind, from which they like to shelter.

Ellis and Ashall noticed that the edges of the groups often coincided quite neatly with the edges of the sun patches, suggesting it was the uneven heating of the sand, cooler below the spread of the plants, warmer in between them, that caused these dense assemblies. Body temperature also comes into the picture, discouraging movement at a certain level and increasing it at a higher. Their preferred body temperature seems to be between 35 degrees Centigrade and 40 degrees Centigrade (95-104 degrees Fahrenheit) and when the sun has warmed them up to this, off they go again. This, of course, is an excessively short version of a highly complicated hypothesis. All that can be said positively is that most hoppers resume their march when the heat rises to a certain level and stop it again when it becomes extreme. This naturally happens about mid-day when a perch even only an inch off the sand may be as much as 10 degrees Centigrade (18 degrees Fahrenheit) cooler than at the surface. At such temperatures a hopper will even climb a small stone in search of a little relief. Most, when they can, clamber into bushes to clump themselves in vertical holds exposing the least possible area of their bodies to the sun's rays, while marchers still on the ground pause to lift one foot, then another, off the baking surface. In cool weather the midday roost may not happen at all; on the other hand, it may last four to five hours. Around sunset the band may begin to group itself around conspicuous plants and bushes and some may roost again, while others, in warm weather, will march on far into the night, a dense, moving, but now very slow carpet of walking insects having no need of moonlight to find their way *en masse* but often seeming confused about their orientation to each other.

In all, such a band will probably have marched for up to a dozen hours, during which its members will have exhibited all manner of behaviour. During the cool hours they will have walked; during the hot they will have hopped. Sometimes they will have marched in bursts of a few seconds, at others remained almost stationary or, at most, pottered around making searching movements with palps and antennae. Those in the first instar may often, during the morning pause, have been clumped, sitting on top of each other, a couple of thousand to the square foot, with much coming and going and endless jumping and fidgeting. Now and again, as if trying to get a bearing on some near object, individuals or groups of hoppers will be seen at a standstill, their forelegs extended, the foreparts of their bodies raised as they move from side to side. Yet overtopping all these many activities there is one common impulse, to move and to cohere.

There is something relentless about this onward movement, as though it had some positive purpose. And yet there is no apparent purpose, except to survive into adulthood, to mature and multiply, to swarm and breed and again pursue the same cycle of life, always as a mass. It is as though a mindless life force were animating all these millions of moving creatures, propelling them on towards a collective life which, as soon as they emerge from the bush or desert into the cultivated lands of man, must be totally destructive. Once they have begun to come together, every instinct and characteristic seems to work in a powerful attempt to hold them together. This may not be successful, for there are fortunately even more powerful forces which sometimes work against them, so that the bands are destroyed or scattered before they fly, or, having flown, are diminished and dispersed. But these are, as it were, the casual forces of nature – droughts, predators, killing winds – whereas the locust's own forces are in-built, persistent and inter-locking, and are always directed to survival under the most adverse conditions, in the individual and the mass.

In the interests of the band or swarm, moreover, the individual will always be sacrificed. Cannibalism, common enough in the insect world, becomes, in this context, a way in which a hopper band keeps going, maintaining its march even though the rains may have been scanty and the food supplies of vegetation consequently thin. We have seen, also, how by travelling in great numbers the insect is able to maintain a cohesive movement for days on end even though its predators are able to gorge and gorge again. In ways not yet confirmed by the laboratory, it may also protect itself against the searing heat of sun and wind by its behaviour during the mid-day roosts, in which the ever-shifting masses seem to create their own micro-habitat, a little temporary shelter composed of their own bodies, each of which in turn screens the other; just as, in the same way, by huddling, they may protect each other from extremes of cold.

In a creature which is *always* gregarious, much of this would be readily understood. We take it for granted that deer or wild cattle will herd together, that gulls will flock, that ants will swarm; but in the locust, as Uvarov proved, this is not necessarily so. How, then, does this enormous cohesive force arise, to reverse and over-master almost every characteristic of *Schistocerca* in its solitary phase? How can any part of its behaviour pattern be inherited when there are these abrupt breaks, occurring unpredictably at intervals of years?

The answer, surprising as it may seem, is that no locust is born with the ability to group but must acquire it afresh in every genera-tion. This is simple enough for the progeny of parents which have already become gregarious. Eggs are laid in groups and from the moment of hatching the young hoppers find themselves so crowded that they are conditioned into gregarious behaviour very quickly. *Solitaria* hoppers may also learn to group under various circumstances. The habitat in which they normally live their quiet, unharmful lives may be invaded by swarms or hopper bands. Contact will then enforce the learning of gregarious ways and if continued will produce a complete phase change. Environmental conditions – the winds, state of the soil, availability of food and shelter and other natural factors, all of which we will discuss later – all contribute to this complex process.

However the initial gathering of the parent locusts may have occurred, whether they may have been blown together by cyclonic winds sweeping over a vast area or herded by contracting vegeta-tion as the always-brief greenery of the wadis has withered in the presence of drought, or for other reasons still the subject of study, the result is that their progeny are no longer free to shun each other as during the solitary phase. It is no use for one hopper to try to shift out of sight of another, for under these conditions his longest leap is only likely to land him on top of a third. Forced to acknowledge each other's society, they begin to inspect, touch and feel each other. What in human terms would be a calm, contem-plative life becomes a nervous and excitable one. The solitary insect which previously has done little more than sit in a bush, moving only to sun itself or feed from time to time, is now brought face to face with neighbours. A great deal of antennae pointing and whirling goes on when, on meeting, each endeavours to examine the other. Touched on the abdomen, the insect kicks out vigorously and in doing so may disturb others which promptly join the activity. These reactions are all part of the learning process by which individual hoppers are conditioned to the group. They may have started life in the uniform of black which shows that they have descended from gregarious parents, but they must still go through this mutual schooling in order to establish their own gregarious-ness. It can all be done quite quickly – in about four hours as observed in the laboratory – and, in the case of some locust species, can be unlearned almost as fast, so that an insect deprived of its fellows' society returns with little delay to all the more evident characteristics of the solitary phase. The Desert Locusts, having a harder row to hoe than others, have greater need of cohesion and, perhaps because of this, are slower to discard their new ways once they have acquired them. So long as they are together, they will constantly be inspecting and touching each other, establishing and maintaining the mutual attraction which is the basis of band or swarm.

There is, it seems, a definite pathway of change along which the locust may travel or return, going all the way or only part of it. At each end of the path there is a pattern of behaviour with which it must accord. In the middle the pattern will be less clear. When it reaches it the locust is in a period of transition. Depending on circumstances, it may go forward or back from it or indeed remain, as it were, oscillating, neither going forward to the completion of one phase nor returning to the other.

As it travels along the path one way or the other the locust flags its change by the colour of its cuticle. As we saw when describing the theory of the phase change, this will at one, the solitary, end of the path, be cryptic or quite green, with hardly any black at all; at the other, gregarious, extreme, the adult will show a marked black pattern on a cream or bright yellow background. Depending on the degree of their mutual inter-attraction, so will the locusts tend toward green or yellow. In their offspring, at the hatchling stage, the extremes will be green (solitarious) or black (gregarious). Thus, at this first stage their colour is controlled mainly by what has happened to their parents; but when they come to the next moult the most important factor is what has happened to them as hatchlings. If they have been with other locust hoppers they either retain the black pattern or, if they were born green, they acquire it; and this development goes on throughout the hopper life, always being governed by their activities or the lack of them. Thus, if they cease to be with others, the green colour returns and the black pattern disappears.

What internal mechanism in the locust brings about these changes is a wide-open question, except that the secretion of certain hormones appears to play a part – as they may prove to do, also, in aspects of behaviour. The necessity for the locust to change its colour when changing from one phase to another is also puzzling, though here one may guess that protection is in part involved, at least in the case of solitaries. But how, when it comes to the black and yellow of the gregarious ones, can the same consideration apply? Monkeys, confronted by insects with strong warning markings, have been known to gibber with fright, and one might suppose that the first sight of a moving mass of black and yellow would temporarily put off birds, but there is no real evidence that it does so and much, as described above, to suggest that it does not – that, indeed, it may work the other way and attract them. Even a very small initial deterrence, however, could at certain times be vital in enabling the bands to build up to a point where subsequent bird attacks could not prevent a plague.

Much more likely is it that the colour change is designed to enable a locust to see another locust so that, once swarmed, they can maintain position and cohesion. This is Dr Ellis's view. In an experiment in a cage, when locusts were separated from each other by a perforated screen, unable to touch but able to see their fellows on the opposite side, the colour change still took place, and was evidently operated by the sense organ of the eye feeding into the nervous system and thus triggering off the required secretion of the hormones. Yet this alone would be a very small part in the whole wonderful process by which the mild, solitary creature of the recession period is translated into the all-devouring collective monster of the plague.

11 A Moroccan farmer ploughs on through a storm of swarming Desert Locusts, which in 1954 destroyed £4½ million worth of crops in one valley alone.

12 '...they ate all the plants of the land, and the fruit of all the trees' (*Exodus*). A field of maize after a swarm has passed. (Inset) Each insect eats its own weight of vegetation every day. A large swarm has been computed to weigh.

Chapter 10

The Plague Returns

If it had been possible to calculate the number of locusts seen in all the national and international surveys made in 1966 it is doubtful if they would have averaged out at more than one per square mile. From the Western Sahara to Rajasthan it seemed as if they had vanished from the face of the earth or had, at any rate, retreated to hiding places impossible to reach. In every district where they might have been expected to breed the sun shone inexorably. If there was rain anywhere in the first ten months of that year it could have been no more than a slight precipitation, insufficient to quench the thirst of the wadi trees and shrubs or leave the needed depth of life-giving moisture enabling the locusts' eggs to germinate in quantity. Here and there in remote wadis a browsing camel might set a few solitaries leaping and winging from tuft to tuft. The occasional flash of wings might attract the attention of some nomad hopeful of an addition to his diet. But report after report was blank of sightings. Nevertheless the recession was nearing its end. Rains, some of them quite heavy, began to fall late in that year and continued, at intervals, into the following summer. It was in the Sahara, early in 1967, that the first uneasy signs of a change appeared.

From spring onwards the monthly summaries of the Desert Locust Information Service began noting reports of small numbers of locusts being spotted here and there in the deserts and semi-deserts stretching for more than 2,000 miles from southern

Morocco to Chad. But it was not until around the middle of the year that news came from Algeria of a flare-up that must have begun in the heart of the desert about March after a burst of rain in the Hoggars. Tamanrasset, so long dry, received two or three exceptionally early and heavy drenchings in the course of the summer. By early June there were 'high densities' at Silet, the southernmost of the Algerian oases, where we had dined so well with Bennounou; 'scattered mature locusts' in mid-June in In Tedeini, 'The Well of the Ticks', and, a month later, 'immature' ones at two of our camping places, Wadi Edjebel and Wadi Ilehr.

Locust people are a unique community intensely interested in each other's personal activities even though separated by thousands of miles. One knew, therefore, than when groups of adult locusts numbering up to a thousand per acacia tree and hundreds per plant had been discovered in June in the wastes north-west of the Hoggars it was Pasquier, camping out again in his beloved desert, who had supervised the count, and his old pupil and acolyte Hadj Benzaza, travelling with him, who organized the control operations aided by soldiers from the barracks at In Salah. Some, probably the majority, of these locusts, were undoubtedly descendants of the scattered solitary ones we had seen in the Wadi In Attenkarer at the southward end of our journey together in the previous year, and of others outside Djanet near the Libyan border, while still others may have originated in the wastes in the southern lee of the Tibesti mountains in Chad. Even when nature, through various causes, takes a heavy toll of hoppers between the time when they emerge from the eggs and the time, up to, say, a month to thirty-five days later, when they should be fledging, the survivors may well be around fifteen times as many as the parent generation; and it was evident that this sort of increase was going on over quite large areas of the central and southern Sahara, wherever rain produced greenery. Control obviously was also becoming more difficult in a country where some of the sites of the flare-ups were separated by a thousand miles or more. In a sense, locust upsurges on this scale are rather like a series of bush fires. While the fighters are endeavouring to extinguish one, another breaks out elsewhere. But a fire at least produces its tell-tale smoke.

The locusts were increasing under cover of shrubs and grass clumps and many of their locations were not discovered at all.

The problem of keeping the locusts' numbers down in such terrain is also hugely increased by the difficulty, sometimes amounting to an impossibility, of sending the available spraying planes, which have a flying radius of at most a hundred miles, anywhere near the scene of the trouble in time to quell it before the locusts have moved on. The wonder is not that the Algerians failed partially in their tremendous task but that they did so well. In spite of all their efforts the result of the breeding that went on practically continuously in the middle and south of the Sahara from February until June was almost certainly a substantial increase in the numbers which were able to wing their way southward over the dunes and rocks of the invisible Algerian border into the twin black desert republics of Niger and Mali. Nor was Algeria the only source, for there had also been the breeding in the Tibesti and the remote wadis of the Fezzan spreading through south-west Libya across the Algerian border. By one of those fortunate chances that turn out to be milestones, the subsequent movements of these insects were observed by a couple of pairs of exceptionally well-trained eyes. They belonged to George Popov and his colleague, also a member of the Anti-Locust Research Centre, Jeremy Roffey.

The two men had been sent to Africa to do field research work at the invitation of the West African international anti-locust organization, OCLALAV. The full title of this body is Organisation Commune de Lutte Antiacridienne et de Lutte Antiaviaire—the 'antiaviaire' referring to the locusts' fellow-pest, the quelea, or weaver bird. Haunting the savannahs in vast flocks, this bird is capable of doing great damage across the breadth of Africa. Most of the former French West African countries supporting OCLALAV are unwilling hosts to both pests. Fortunately they call for a basically similar control organization employing ground teams and light aircraft. This has been built up by OCLALAV, helped by generous French grants, over a number of years, and Popov and Roffey were therefore well equipped to carry out an exhaustive survey not only of those parts of the desert where locusts were known to be present but of many locations, showing

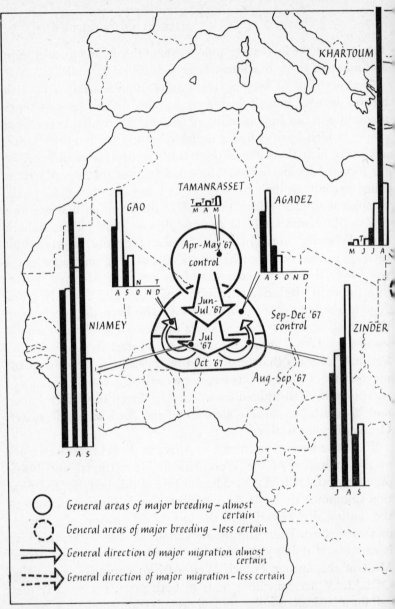

KHARTOUM

TAMANRASSET

GAO

AGADEZ

Apr-May '67
control

Jun-
Jul '67

Sep-Dec '67
control

NIAMEY

Jul
'67

Oct '67

ZINDER

Aug-Sep '67

J A S

J A S

○ General areas of major breeding ~ almost
certain

○ General areas of major breeding ~ less certain

→ General direction of major migration almost
certain

⇢ General direction of major migration ~ less certain

6 *a*, *b* and *c* Rainfall, Migrations and Breeding, November 1966–August 1968.
Outbreaks of the Desert Locust require a sequence of good rains. In 1966 and
1967 these occurred in two main areas – Saharan and sub-Saharan West
Africa and southern and south-eastern Arabia. Successful breeding there was
followed by migrations to both west and east, the latter being the most extensive.
Intensive control operations coordinated by the Food and Agriculture Organi-

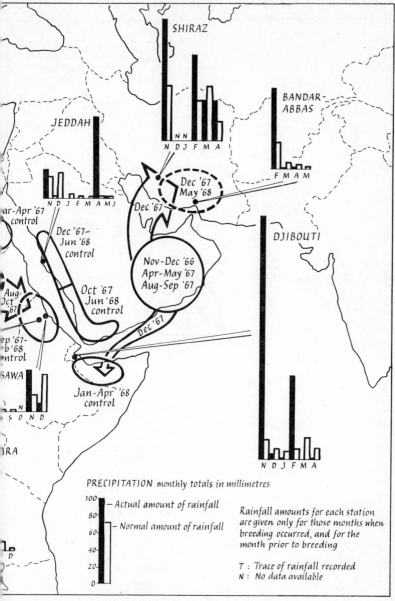

SHIRAZ

BANDAR-
ABBAS

JEDDAH

N N

N D J F M A

F M A M

ar-Apr '67
control

Dec '67

Dec '67
May '68

Dec '67–
Jun '68
control

DJIBOUTI

Aug-
Oct
'67

Oct '67
Jun '68
control

Nov-Dec '66
Apr-May '67
Aug-Sep '67

ep '67–
b '68
ntrol

Dec '67

SAWA

Jan-Apr '68
control

S O N D

N D J F M A

RA

PRECIPITATION *monthly totals in millimetres*

100
80
60
40
20
0

— Actual amount of rainfall

— Normal amount of rainfall

*Rainfall amounts for each station
are given only for those months when
breeding occurred, and for the
month prior to breeding*

T : *Trace of rainfall recorded*
N : *No data available*

zation and supported by the United Nations Development Programme were
required in some thirteen countries. In Pakistan and India the failure of the
monsoon aided human control and the swarms died out. Probably the most
important aspect of the international campaign was that it prevented the
conjunction of the eastern and western outbreaks. (Maps 6*b* and 6*c*, are on
following pages.)

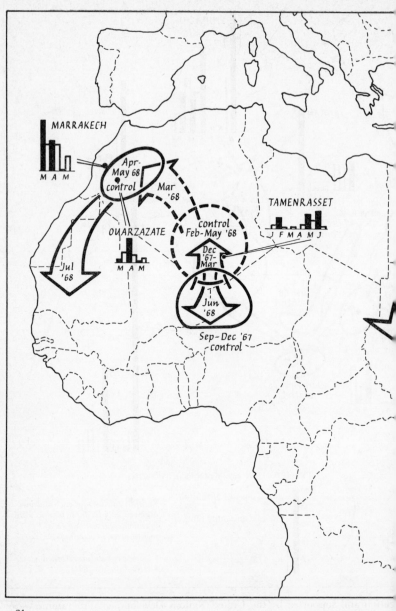

MARRAKECH

Apr-
May 68
control

Mar
'68

TAMENRASSET

control
Feb-May '68

OUARZAZATE

Dec
'67-
Mar

Jul
'68

Jun
'68

Sep-Dec '67
control

6b

up as streaks and blobs of green, where vegetation might be
sheltering others. They were also, in their personal and professional qualities, a very well-suited pair for a work of discovery

136

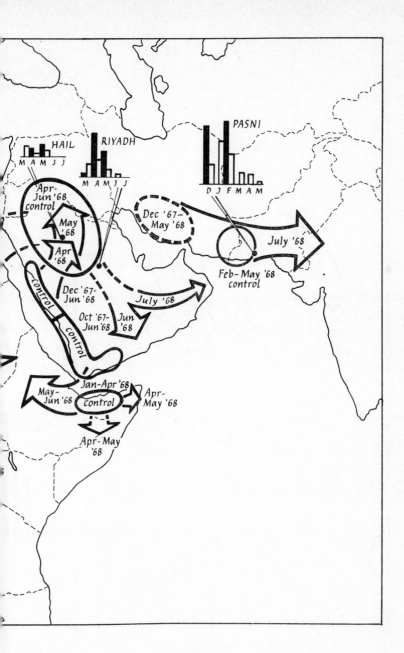

which was to turn out to be a highly important landmark in the study of locust behaviour in the wild on which all effective control must ultimately be based. Both men have had great experience

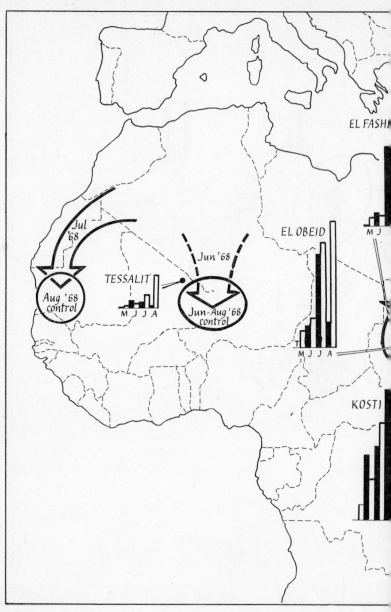

6c

among the insects' desert habitats, both approach their work with
the humanitarian urge – pity for the afflicted, poverty-line farmer –
which is a necessary spur to enduring an often uncomfortable

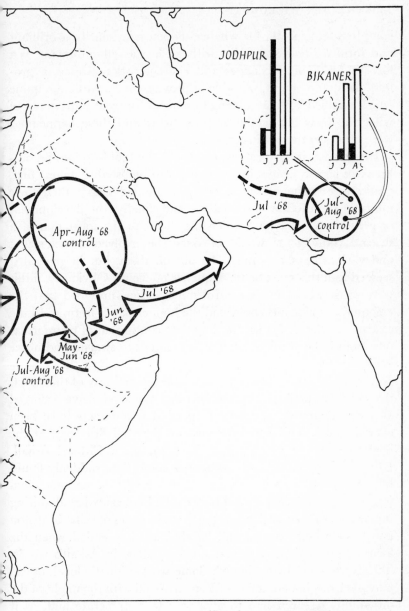

JODHPUR

BIKANER

J J A

J J A

Apr-Aug '68
control

Jul '68

Jul-
Aug '68
control

Jul '68

Jun
'68

May-
Jun '68

Jul-Aug '68
control

(See also maps 8*a* and 8*b*, pp. 172-3.)

life, and both are good linguists, able to get on with the nomad
and semi-nomad people who can be a valuable source of information. They also, in this particular instance, were able to use

complementary skills, for whereas Popov is primarily a brilliant
and intuitive field naturalist with an eye as all-embracing as a
locust's for habitat changes and characteristics, Roffey, a professional entomologist, was able to provide the scientific discipline
which enabled both, as a team, to make a coherent pattern out of
what they saw happening around them; and these happenings
were often very confusing indeed.

In the past, it has been assumed that the build-up into an outbreak of locusts occurs in successive, distinguishable stages: first,
the concentration of the insects into limited areas; then their
multiplication following breeding; and, finally, the development
of gregarious behaviour which may lead to the formation of a
swarm. Many factors would, of course, be involved in each stage;
and various emphases might be put on them, but in general it
seemed as if this must be the sequence of events. Obviously, if this
were so, it would make control much easier, provided that the
reconnaissance scouts could find the hopper bands in time. In the
case of the Desert Locust the complete continuum of transformation from isolated solitary-living individuals to swarms had never
been observed by anyone from start to finish. Given the remote-
ness of the possible breeding places and the mobility of the insect,
this is not surprising. Many outbreaks, moreover, have occurred
at such apparently devastating speed that observers have been
overwhelmed. This had happened on the Red Sea coast at the
beginning of the great plague in 1949, when Clifford Ashall,
doing some research work on hopper coloration, suddenly found
his camp awash in a flood following a rainfall of seven inches in a
single night – this in a district where it had not rained at all during
the previous five years! It was the first of a series of rains falling on
and off for four months and barely had they ended when the
locusts appeared in such vast numbers that, in Ashall's words,
'they were going by for hours'. Some were evidently local, others
may have originated in far-off Yemen; faced with the might of the
onslaught, research had to give way to a desperate and, as it
proved, unavailing battle to prevent the plague from spreading
and those on the spot could only conjecture the place of the swarm's
origin, though it is now thought certain that it must have been in
the interior of South Arabia.

Popov, with his vast experience, had himself seen only parts of a build-up, and these in various places. Discussing his experiences with me, he said: 'I had seen the processes from, so to speak, A to D, F to M and Q to Z, but never the whole alphabet. It was the same with Roffey. We had our ideas of what happened and in the Tamesna of Niger we had the marvellous experience of seeing the whole process unfold under our eyes.'

The two scientists arrived in Niger in September 1967 and made their forward bases at In Abangharit in the Tamesna, 240 miles south-east of the point I had reached with Pasquier, and at Termit Sud, 250 miles north-east of the OCLALAV base at Zinder. The Tamesna, nearly all desert, consists of a vast ancient flood plain where normal rainfall is relatively light. Periodically, however, storms break over the two principal massifs, Aïr and Termit, causing flash floods enabling ephemeral plant life to spring up in wadis and depressions many miles distant.

This was the situation when Roffey and Popov began their observations following a spell of unusually good summer rains similar to those which had been falling in the central Sahara. Tamanrasset, on its spur of the Hoggars, had had a downpour of more than an inch in July. The resulting floods had reached Tit and Silet, the two oases on the track leading southward toward the borders of Niger and Mali, and heavy breeding with dense concentrations of hoppers had been seen in both places. There were also reports of scattered mature locusts at various points along the caravan trail. All this accorded with the Desert Locust Information Service's forecast, issued from London, that the Saharan locusts would begin moving south across the Algerian border. In fact, a 'very small' swarm of grey and brick-red locusts had already been seen in Niger on 25 June, when they settled over a hundred or so acres before moving on south-eastward. By early August wadis of the Adras des Iforas, a rocky massif in north-east Mali, were infested with hoppers clustered forty to fifty to the tuft. During the first eleven days of August just over $1\frac{1}{2}$ inches of rain fell at Tessalit in the north-west corner of the Adrar in a zone where half that amount in a year is normal, and when Popov flew over the Tamesna soon after his arrival the conditions for the invaders were ideal over much of the area.

That Desert Locusts in some numbers had been displacing southwards was proved in mid-August when an OCLALAV team exploring in the neighbourhood of the Adrar before the Britons' arrival found clumps of tribulus alive with the yellow-and-black bodies of a considerable colony of gregarious early hoppers. Fully-fledged locusts in the grey-pink stage of early maturity were also discovered along with a number of old locusts which may have fallen out of a swarm. In the main, though, most of the insects seem to have been flying individually – only one swarmlet of a few hundred locusts had been seen, in Niger.

This discovery was the signal for an important little dusting and spraying operation which killed most of the insects and undoubtedly checked a local outbreak. Such a measure, however, can seldom be fully successful when the weather is in favour of the locust, as it now all too obviously was, not only over a large part of the Sahara but in desert country stretching far to the east and to the west. A small proportion of the enemy are bound to escape and like the soldiers of a defeated guerilla army are liable to scatter only to form the nuclei of fresh detachments elsewhere. And there must also, of course, have been other groups of locusts, more or less homogenous, which were not discovered. Gradually, therefore, their forces began to build up and by the end of the month scattered locusts were being reported from a number of points in Mali and Niger territory, including the wadis of the semi-mountainous massifs, Aïr and Termit, standing up like islands amid the vast seas of sand-dunes and arid plains stretching south-eastward toward Lake Chad. Both of these had been indicated as possible danger points by Popov, when reporting on his ecological survey, for like other substantial geographical features whose pronounced relief makes them so grateful to the eye of the traveller wearied of the huge monotony of the surrounding desert, these lonely rock massifs tend to have their own sub-climates, miniature weather systems and mosaics of soils all tending to produce a wider-than-normal range of plant communities in the kind of patchwork favourable for an increase.

Thinking over the evidence of their own observations later, however, Roffey and Popov concluded that most of the locusts escaping Algerian control must have overflown the Tamesna, which was then still largely dry, and continued southward until they reached the so-called Sahelian zone some 100-200 miles beyond. This long and relatively narrow belt of country begins roughly north of the bend of the Niger at Timbuktu and continues through Agadez to the northern shores of Lake Chad. Its name derives from the Arabic Sahel, implying sown land, and its chief charms for the Desert Locust are its rainfall and its patchy but fairly stable vegetation, including some primitive cultivations of millet. The rainfall is meagre – normally only 150 mm to 200 mm (5¾ inches to 7¾ inches) a year. Nevertheless it is usually enough to ensure sufficient soil moisture for successful egg-laying from July through to September or even longer, and, of course, it also creates the hoppers' needful habitat.

All these circumstances were in the locusts' favour in the late summer of 1967 and must have enabled them to increase their numbers tenfold or even fifteenfold. The two watchers and their helpers, camping out in the Tamesna, were not surprised, therefore, when toward the end of September the beams of their Aldis lamps picked out the fluttering shapes of more and more invaders flying toward the north. The majority of them were a new generation which must have been born in the Sahel and their ever-increasing numbers warned of still more to come. By this time rains had come to large parts of Tamesna and conditions there were in every respect right for the critical third generation.

Seeds which had lain dormant in the same place for years or had been blown along by the searing yellow winds until finding lodgement in the sand-wicks criss-crossing depressions in the desiccated clay plain had germinated, mantling the wilderness in green. Schouwia, the same blue-flowered annual I had seen growing so thickly in Wadi In Attenkarer the year before, rapidly put up a waist-high growth in which the incoming insects throve. Tribulus, another of their favourite food-and-shelter plants, was also shooting healthily when Roffey and Popov arrived to find the first notable signs of what was evidently a major upsurge.

There are, as remarked, two theories about the causation of locust plagues – the first, that they are carried over from one period to another by a swarm or swarms which are always in being somewhere or other, and in some form or another, in the five million or so square miles of potential breeding ground (in other words, there is never a complete recession, only a lull); and the other, that the locust starts each time from scratch as a solitary recluse but builds up inordinately when the appropriate conditions arrive. These basic requirements are three successive rain periods (with accompanying vegetation) spaced out in time and distance permitting the birth of three generations able to migrate readily from one rain area to another, increasing as they go. Allowing for destruction by predators, and other natural causes, each generation can be reckoned to increase at least ten-fold. That is to say, given the right amount of rain at the right places at the right times, the locusts can within a year or eighteen months multiply their numbers a thousand times, and a plague may begin.

For reasons to be seen, this may be called the Vegetation Theory, and Roffey and Popov had long been strong proponents of it. As presented above, it might be assumed that the whole process would be more or less clear-cut. In the event, had the locusts set out to confuse the picture they could hardly have been more successful. Some had copulated and laid their eggs to hatch out before the men had set up camp and others, in various stages of maturity, were coming all the time they were there. They were therefore surrounded by insects of all ages. These had to be sought and studied by day while arrivals continued by night.

In a report on their field work and conclusions, published in *Nature*, Roffey and Popov have defined the fundamental processes in the build-up of a locust outbreak, as they saw it, as *concentration*, *multiplication* and *gregarization* and they underline the fact that here amid the harsh open spaces of the Tamesna (and presumably wherever else the Desert Locust had begun to pose a threat), all three processes were happening at once, so that even in a relatively small area locusts were to be found in every stage of the individual life-cycle, at every scale of density from the most thinly scattered to the most tightly packed and showing every aspect and degree of that condition of togetherness which is what is meant by gregarization.

So thorough was the mix-up, so interrelated and overlapping were the processes, that there were times when it was difficult to ɔe sure which processes were operating at any one moment on ɪndividuals or groups. Yet little by little their patience was ɪewarded and an underlying pattern began to show. It became ɪpparent, for instance, that the prime factors causing the insects ɪo concentrate were, first, the distribution of the general habitat ˈanticipated by Pasquier twenty-five years before) and, second, ɪheir preferences for some parts of it rather than others. Although ɪhe Tamesna at this time had an exceptional amount of vegetation, ɪhis in fact covered less that ten per cent of the total land surface. ɪΓhe locusts' choice of landing sites was therefore narrowed ɪccordingly and an initial degree of concentration naturally en-ɪued. Then, as they scrambled and flitted around seeking out their ɔreferred food-and-shelter plants, and suitable soil for laying, ɔockets of still higher concentration began to occur.

Making systematic traverses by foot and vehicle and counting ɪhe locusts flushed, the British observers estimated that by mid-ɔctober there were about five million adult locusts in the Tamesna ɔf Niger alone. Had these been spread out equally over the whole ɪrea they would have averaged about one locust per $1\frac{1}{4}$ acres, ɪwhereas they were in fact packing themselves into favoured ɪabitats at from fifty to a thousand times the average density.

While all this was going on more and more adults were coming ɪn, increasing the excitement of those already present. The upsurge ɪn their sexual activity was particularly noticeable. Normally ɪndifferent male solitaries, in whom the urge to copulate seems ɪlmost a matter of chance depending on the propinquity of a female, became restive and aggressive, ready to fight for her possession. Γhe presence of adults speeded up the sexual maturation of others ɪand while this was happening their behaviour changed in other ɪways. They began to crawl about more actively and make daytime ɪlights, all of which led to further concentration. The daytime ɪlights, moreover, were markedly different from those of the night. ɪn fact, flight seemed almost an exaggerated word for a typical ɪingle movement covering only twenty metres or so. Such move-ɪments, however, usually occurred in rapid succession. Ten or ɪwenty low, short skimming flights ended as a rule in an open ɪpatch between the vegetation and if a male and female found ɪthemselves there together they coupled. If the purpose of the males'

flights and crawling was largely sexual, as this suggested, that o
the females enabled them to discover suitable laying sites. Eithe
their flights brought them into the proximity of other female
already laying, whom they joined; or they would find a site which
had already been used, recognizing it probably first by smell and
touch and eventually by digging; or they would hunt around and
find their own site. But in no cases did they lay until satisfied tha
the conditions required – adequate but not excessive moisture and
loose-textured soil – were right.

This was only part of the multiplication process. In actuality
once the maturing stage was over, the females had to search
industriously, even under these near-ideal conditions, to find suitable laying spots. By mid-October most of the area had dried to a
depth of five to six inches and it was only here and there, in bare
patches among the schouwia and tribulus, where the sandy cracks
retained more moisture, that the delicate sensory mechanisms o
their extended ovipositors, pressed downward into the sand, sen
back the message that all was well. In one such spot Popov and
Roffey dug up 71 eggpods in approximately one square foot. Had
they been able to hatch out – and elsewhere most of them did – the
result would have been some 7,000 little hoppers all aspiring to
become locusts. Multiplication, it seemed, could hardly go further
and fortunately a natural mechanism was at hand to see that the
rate was seriously cut down. From the stage of being an egg to the
stage of being a fully-grown locust is a hazardous progress at
the best of times, and undoubtedly these *were* the best of times.
Nowhere in the egg-fields were there signs of parasites and nowhere
in the habitat, faintly stirred by the slightest of evening breezes,
were there egg predators, no flies, beetles, ants or wasps. Even the
predators of hoppers and adults, birds, snakes, lizards and rats,
were uncommon in this seeming locust-Eden. Yet ninety-two out
of every hundred hatchlings died somewhere along the path of
growth between the first and the last instars. I have suggested
elsewhere, on no scientific authority, that once the processes of
concentration and multiplication have begun, nature takes care
of *all* contingencies and hazards at first by over-producing and
then by pruning. Had the Desert Locusts' natural enemies been
on the scene in normal numbers the mortality rate would con-

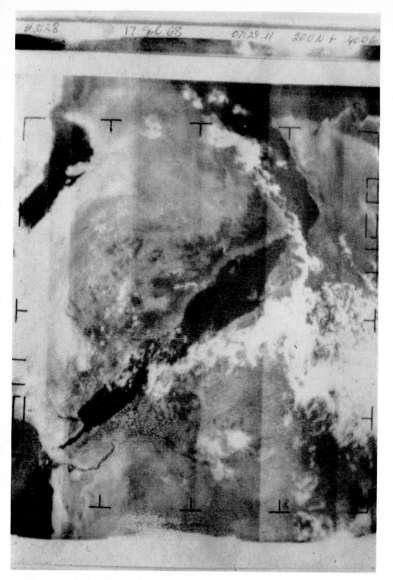

13 Experts of the Desert Locust Control Organization of Eastern Africa gain valuable information as to the possible whereabouts of swarms by daily studying the position of the weather front as photographed by a satellite sending signals to the US Air Force tracking station at Asmara on the Ethiopian highlands. The cloudline in this photo shows the front extending across the foot of the Red Sea from southern Arabia into Eritrea and Sudan. Suez is in the south-west and the Persian Gulf in the north-east.

14 Swarms were formerly treated with sodium arsenate dust. The kill was minimal but gave needful employment to subsistence farmers whose crops had been ruined.

ceivably have been higher. But they were not there. Either the environment was too adverse or the larder too chancy for any obligate species of locust-enemy to be able to establish itself. So nature used some other, unidentified means and killed off the surplus just the same. The watching scientists made a rough calculation that in the southern Tamesna the five million immigrants which had arrived in solitary night-flights had laid 750 million eggs, most of which hatched out successfully; and that by the middle of October at least eighty million of the hatchlings had survived to the fourth instar. At this stage the hopper is difficult to kill. The multiplication rate was therefore about sixteen times. There was no doubt in the watchers' minds that they were witnessing the beginnings of a very serious outbreak – a panoramic view, in one place, of all the events which may normally have preceded the outbreaks of any of the great plagues of history.

Rains, the growth of vegetation, the gradual concentration of scattered locusts, mating, breeding, limited migration within the habitat, further concentration as the areas of vegetation dried and shrank, all the phases of gregarization from the marked changes in body colour to the beginnings of morphological change, the banding together and marching of the hoppers, successful fledging and putative swarm formation – all were part of a classic but previously only partly recorded picture.

Within it there were many fascinating brush-strokes. On one occasion when the party's Landrover had been driven across a patch of sand where the average moisture level was about four inches below the surface, too deep for successful egg-laying, the wheel-tracks were being used as laying sites within three or four hours. Biting into the sand, the tyres had reduced the average depth of dry sand to about two-and-a half inches. Below this there was moisture and the females had found it. Their urge for togetherness, moreover, had caused them to gather in groups. In one group Popov counted twenty-nine industrious mothers crowding a short section of wheel track into which they were thrusting their ovipositors with every satisfaction. Various reasons have been advanced to explain the female locusts' disposition to get together when laying. One is that they are attracted by the odours left by others in the egg-field; another, complementary to this, is that

L

once the first few females have loosened the ground, capillary action in the soil and egg-pod froth raises the moisture level and so results in an over-all improvement in laying conditions. Whatever the cause, the result is dramatic, for out of this small patch of soil will emerge the greatest possible concentration of hoppers and all of them will be behaving gregariously within hours of their emergence from the pods. Soon, when a number of these pod clusters have hatched out close together, the ground will be covered with a shimmering mist of their tiny leaping bodies.

This, and their later transformation into marching hordes surging forward over the sands with their leaders tightly massed along the front and the rest scurrying to keep up, is one of the most ominous and chilling sights to be seen when locust-hunting; and in the Tamesna, before control got under way, there were many such bands. At several points, when hatching was at its peak in mid-October, Roffey and Popov counted densities of up to five hundred to the square metre. Elsewhere they might be thinner, as few as a couple to one large tribulus plant, and where this was so they kept to their solitary ways, avoiding one another as far as possible and doing their best to hide when an observer approached. The emergence of new hoppers from later pods coincided with the shrinking of the vegetation as the ground dried out and this increased their density and caused changes in behaviour and colour, for they then began to encounter each other willy-nilly, thus beginning the conditioning process which would eventually induce them to stay together and become increasingly gregarious. This showed itself in group formation. Another sign of gregarization was their action when disturbed. Instead of staying in the roost and hiding in the bush when intruders were around they would fountain out of it, leaping and hopping across bare ground to seek refuge in another plant which might be several yards away. As one group of fugitives fled, others in other bushes would begin to do the same so that for twenty yards or more from the original point of disturbance there would be a rippling wave of leaping locust hoppers all likely to end up closer than before.

No one has previously considered disturbance as a factor promoting the gregarization process, but, witnessing this behaviour, Roffey and Popov concluded that in the Tamesna, at least, it is an important one. For on these great plains it is not only locusts which profit by the rains and enjoy the sweetness of the herbage. Camels arrive too, sometimes in great herds with a Tuareg boy in charge. The ability of the Tuaregs and other nomads to sniff out rainfall and the ensuing vegetation is one of the mysteries of the desert, for like locusts they arrive from long distances away almost as soon as the ephemeral pastures appear and stay while there is any grazing remaining. Antelopes, too, are attracted to these rare lush feeding grounds and, as I had seen, much farther north in the Sahara, even a cattle-drover may appear, driving his thin beasts from God knows where to share the fleeting bounty. When all are present at the same time all interact together. The camel is not averse to munching up a few locusts with the leafage, the nomad catches them to produce a nice delicacy over the evening fire and the disturbed hoppers, leaping away together, become ever more gregarious.

Desert Locusts were not the only members of the grasshopper family infesting the Tamesna that autumn. In one locality there were at least three other species about ten times as abundant. Finding itself among strangers, however, produced the same effect on the Desert Locust hopper as when it met with numbers of its own kind – its body colour changed, showing black markings, indicating the beginnings of gregarization.

Having gregarized, the late-instar hoppers formed roosting, basking and feeding groups, and this they would continue to do for the first few days after casting their last cuticle to shake out their wings as fledglings. In their first flights some gathered in acacia trees and quickly formed swarmlets, needing urgent action by the OCLALAV spraying team with whom Roffey and Popov were working. Those which got away did so at night, flying off in the same state as their parents had arrived, as solitary individuals. Comparatively few of the original immigrants having gregarized, it was this succeeding generation which gave the scientists their most important evidence about the way increases in numbers and their distribution in the vegetation affect behaviour. In general the parental population had to number about 250 to the acre before enforced mutual acquaintance persuaded them to abandon their solitary ways and settle down together in basking and roosting groups. In this state the males' sexual liveliness was also noticeably stimulated and they would start searching around for ready females, who after copulation would form their own groups for egg-laying. The observers also remarked that the greater the concentrations became the greater proportionately were the numbers of males sporting the bright yellow colours of gregarious maturity. The greater their gregariousness, as shown by behaviour and colour, the greater, moreover, was the degree of these qualities among their progeny. At Mkibas, about sixty miles north-east of In Abangharit, where the hatchlings numbered hundreds per square yard, they began forming basking groups immediately after shedding their papery skins or emerging from the pod, while elsewhere, at lower densities, this would not happen for hours or even days. In the same place the influence of the parents' advanced degree of gregarization manifested itself in about a quarter of the hoppers either turning black within three or four hours or showing black markings very soon afterwards – both being signs of inherited gregarious tendencies.

Equally important, for it may very well play a significant part in future control tactics, was the fact observed by Roffey and Popov that the nature and pattern of the vegetation itself had a considerable influence on the extent to which hoppers banded together, and marched. Where it was thin and patchy this happened earlier. By the time they had reached their third, fourth or fifth instars the hoppers in such places, if they had become gregarious, were likely to start marching spontaneously and could easily be found. Where their favourite habitats of tribulus and schouwia were lush and thick their numbers were frequently greater – indeed often very great – but, although they formed into groups, they rarely marched spontaneously. A disturbing intruder might set them all in motion, otherwise they remained concealed.

The danger of this was obvious to the watchers. For a big population like this could lurk unsuspected, poised on the brink of full gregariousness, and then break out on a large scale and with dramatic speed. This happened on two occasions when a dense population of hoppers which had previously been quietly basking, sheltering and feeding, erupted suddenly from their hiding places and within five or six days were in full spontaneous march.

If Roffey and Popov were right in their conclusions it meant that previous local outbreaks, precursors of plagues, may have gone unspotted until too late because no one had looked in the right places. Under cover of a thick habitat the locusts had met, mated and built up their numbers into formidable gregarious cohorts only to be detected – if anyone was around – when they at last burst out.

Sticking out their scientific necks, the two men expressed the opinion in their paper in *Nature* that, in the past, the appearance of conspicuous bands of hoppers may have been attributed wrongly to laying by swarms. The 'may have been' was tactful. It did not directly contradict the theory that somewhere or other a swarm is always in being and that it is through such swarms that one plague is carried over to another. But by demonstrating that this West African plague did start from initially solitary-living locusts it came pretty near to doing so. It also invited the corollary that successful detection requires thorough surveying by people who can recognize not only locusts but also their habitats and behaviour: in a couple of words, locust ecology. The method adopted in the Tamesna combined air and ground survey techniques. First the habitats suitable for locusts were located from the air and then they were inspected on the ground. But there and elsewhere the locusts were too many for the forces then available.

The Saharan outbreak witnessed by Roffey and Popov was, as it turned out, one of five which occurred following exceptional spring rains in west, north-eastern and eastern Africa and Arabia. Most of these had probably originated from undiscovered local populations which had only required a beneficial change in their environment to enable their numbers to multiply. The fifth region to be affected, the Iranian coastal plains of the Persian Gulf, was in one sense the most worrying, for it was almost certainly a true invasion by swarms which had bred in Arabia and crossed the Gulf. Unless stopped, they could well carry the plague into Pakistan and northern India.

The new outbreak in Arabia occurred in the region of the Empty Quarter where the 1948-9 plague had started under very similar conditions, following rainstorms of extraordinary intensity. There was, however, this difference. In 1948 the meteorological origins of the flare-up could only be guessed at a long while afterwards. This time they were witnessed, not only by the region's Bedouin – to whom they meant nothing – but by the all-seeing eye of a camera lens on an American weather satellite. This signalled its photographs on three consecutive days in late 1966 to the Air Force tracking station at Asmara on the Ethiopian highlands on the other side of the Red Sea. Clicking out line by line on the facsimile recorder in an unassuming-looking blue trailer next to the aerial just outside the town, they showed first – on 10 November – the white silhouette of a vast mass of clouds over the Arabian Sea south of Pakistan and east of Muscat and Oman; then the same clouds over the Oman coast; and, finally, in the same position, the ragged shape of the cyclone breaking up. In fact, it was at that time dumping nearly four inches of rain – more than the average for the year – on the Bedouins of Salalah on the coast of Oman and the deserts far inland.

One cyclone, however, is not enough to launch a plague unless followed by good rains at suitable times and places elsewhere. These requirements were fulfilled when exceptionally heavy and widespread rains again swept over southern and eastern Arabia in the spring and summer of 1967. In parts of the Rub al Khali (the Empty Quarter), where normally rains fall only about once in five years, there was the extraordinary sight of lakes of standing water lasting from four to five months. In a subsequent survey Popov judged that within a year the original locust population must have multiplied perhaps a thousand times. Even so, most of them departed as solitaries, flying by night. Reversing Shakespeare's dictum as to sorrows, they travelled not in battalions but as single spies, to swarm elsewhere. Iran, as we have seen, was among the sufferers.

Meantime, the other side of the Arabia Peninsula had also been getting its share of rain, causing a spectacular upsurge of locusts in districts previously thought to be empty. These included the southern Tihamah, the long strip of littoral desert and rock south of Jeddah and Mecca, hemmed on one side by the Red Sea and on the other by the beautiful mountains of Asir. Travelling in this region is, if anything, more difficult and certainly more uncomfortable than in the Sahara. For long distances there is only one track and at one point it takes to the sea to avoid a long stretch of lava boulders blocking the land. Adequate scouting there really requires a combination of air and ground search, the former to spot the green areas indicating recent rainfall, the latter to comb these areas when found. No aircraft were at that time available and the ground teams did the best they could; but by midsummer it was evident that large numbers of Desert Locusts – not yet swarming – had moved into the Tihamah wadis, and by October gregarious hoppers were banding and marching both there and in South Yemen. Since the Red Sea basin is a notorious epicentre for plagues, there could hardly have been a worse place for an outbreak.

In Ethiopia there had been a good deal of well-founded worry ever since the Asmara satellite-tracking station had picked up pictures of the Oman cyclone in the previous November. The territory between the mountainous inland escarpment and the coast is difficult to watch for a variety of reasons and when rain brings floods, as happened toward the end of 1967, much of it is impenetrable. Under cover of the scrub the local locusts multiplied rapidly. Their numbers seem to have been further bolstered by invaders flying in from southern Arabia and when the year ended most of the Eritrean coast, the Territory of the Afar and Issa Peoples (former French Somaliland) and the north coast of Somalia were all in trouble. In Sudan a major flare-up had already begun.

On 27 December 1967, the Director-General of the Food and Agriculture Organization issued a special warning to some thirty countries liable to be affected, that if the population build-up were to be followed by successful breeding in the spring of 1968 a plague could recur. By the spring successful breeding had taken place in at least a third of the endangered territory, and the battle was on.

Chapter 11

The Campaign in Eastern Africa: 1 – Ethiopia

The Desert Locust Information Service summary of the summer situation opened with the laconic statement, 'Swarms are now present in all the major summer breeding areas and breeding is known to have begun in most of them'. The accompanying chart made it look as though a vast game of noughts and crosses was being played right across the map of Africa. Another was in progress in Saudi Arabia and sister Arab nations to the south. Beyond the Persian Gulf it looked as though a game had just ended in Iran while another was starting in Pakistan and North-West India. The crosses (and stars) were the symbols of swarms, white circles were egg-fields and black ones hopper bands. Brackets showed non-swarmers, in groups, isolated or unspecified. Naturally there were some approximations. What the chart demonstrated clearly was that gregarious locusts in greater or smaller numbers were patch-worked across a four-thousand-mile-long band south of the Sahara from the western Atlantic to the Horn of Africa and that there were all the possibilities of this being stretched from the eastern shores of the Red Sea to the plains of Rajasthan.

In eastern Africa some locusts were in the Red Sea plains, some drifting eastward into the north-west corner of Somalia and many were in the difficult, hostile country of the Eritrean highlands, toward the border with Sudan. They were difficult to find and the cut-up nature of the terrain, sodden with more rains, made them difficult to control. In Sudan the situation was becoming increasingly alarming. By the end of July five of that huge country's nine provinces, each as big as a sizeable European country, reported heavy infestations. On the Ethiopian side of the border swarms were known to have been laying in the westerly uplands and along the border district facing the nearest Sudanese city, Kassala. North-west of Kassala the waters of the River Atbara, a major tributary of the Nile, flowed high and yellowed with sediment through a desert landscape newly clothed in green. Swarms were laying at several places along it and also eastward in the triangle between this river and the Middle Nile north of Khartoum. It was obvious that these two regions, the Ethiopian west of the highlands and the Sudanese east of the Nile, were in effect all part of one big infested zone, and with the possibility of fresh arrivals from Arabia more trouble was expected. It was in the countries around the Red Sea and the Gulf of Aden, therefore, that the momentous battles were at this time being fought.

FAO had, fortunately, its own machinery for enlisting common action. For many years past the meetings of its Desert Locust Control Committee had brought together experts representing all the nations affected. The first of a number of important regional groupings, also fostered by FAO, linked India, Pakistan, Iran and Afghanistan. There were two long-standing independent communal Desert Locust control organizations, DLCO-EA and OCLALAV, in eastern and western Africa respectively, enjoying the support of fifteen nations ranged across the continent from Senegal to Tanzania. The constitution of a similar one, planned for the Middle East, awaited general ratification – a result quickly speeded up by the spread of the plague. Action in all cases was to be based on the principle that each member country would be primarily responsible for its own protection, but that all would come to each other's aid if required. FAO's role was to coordinate all the national and regional campaigns in a common strategy for which the wherewithal was provided by the United Nations Development Programme. The most important single factor, however, in making all this paper cooperation work was a human one. FAO's Chief Locust Officer, Gurdas Singh, had spent the whole of his adult life in locust-control work, beginning as an entomologist in his native India and including every campaign afterwards.

He was that rarest of breeds, the man of action who is also a good administrator and diplomatist. It was at his suggestion that I went to eastern Africa to see cooperation in action.

I arrived in Asmara, Ethiopia, at the beginning of July, 1968, as daylight was breaking. There were two of us, the other being a tall, bearded, patriarchal person named Tony Izaaks, a producer of the BBC Horizon television programme who was coming out to look at the Desert Locusts' film possibilities before committing the Corporation to the cost of sending a full-scale unit. As usual when going to Africa it was a night flight, depriving one of all sense of spatial separation between the continents, and indeed as we landed at the little airport it seemed as if we had brought the worst part of Europe's climate with us. A line of angry clouds formed a sinuous front more or less parallel to the immensely long escarpment of cliffs edging the plateau on which the town stands; the air at ground level was so dank and chill that we immediately put on overcoats; rain spattered as we piled our luggage into the DLCO Landrover. Yet most striking of all was the country's greenness, all the more vivid for the patches of unsown red soil. For this was Ethiopia's rainy season, and all the reports were at once proved correct. The rains had been very bountiful and the season was being unusually prolonged. All this had brought joy to the hearts of the farmers until the locusts came. A substantial swarm had swept over the outskirts of Asmara some days before and their immature, dead, pink bodies, smelling heavily of poisonous spray, still lay thick on the ground. It had been a quick, efficient air operation, all the better for the fact that the insects were killed before they could lay and multiply.

The Desert Locust Control Organization of Eastern Africa is the successor to the Desert Locust Survey set up by Uvarov at the beginning of the previous plague when Somalia, Kenya, Tanzania and Uganda had not yet achieved independence. These four countries, with Ethiopia, now pool an important part of their anti-locust resources. Sudan has recently joined and contact is also maintained, either directly or through FAO, with Saudi Arabia and her southern neighbours whose territories were formerly covered by the British survey. Gurdas Singh, when posted in Ethiopia, had played a considerable part in putting DLCO on its

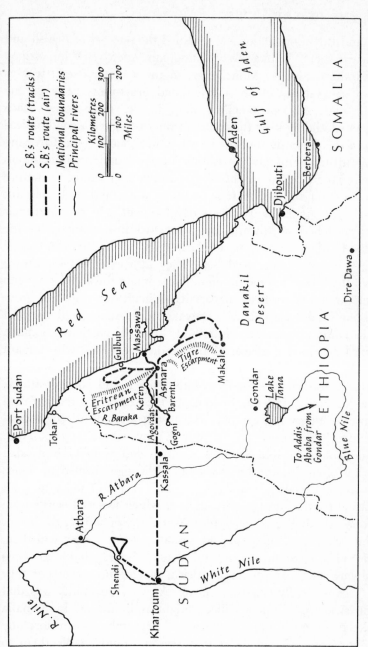

7 Author's routes in Ethiopia and Sudan.

feet and the organization has retained the services of British and
African experts. With a shrewd Ethiopian, Adefris Bellehu, trained
in administration at Manchester, in the director's seat, it is as
good an example of an international grass-roots body as one
could find. A floor-to-ceiling map in the three-storey headquarters building near the centre of the town gave a good indication of the dimensions of the locust control teams' task. Covering
an area the size of western Europe, it was studded with symbols
showing the whereabouts of swarms and hopper bands, the nature
and dates of the action taken against them and such relevant
details as their size and stage of development. The pins and their
flags stretched northward and westward from Asmara to the
Sudan border and were particularly thick on the Eritrean coastal
plains, bordering the Red Sea. They also extended southward
down the line of the mountainous escarpment toward the big gap
where a famed railway makes its difficult way upward from
Djibouti on the coast via Dire Dawa to Addis Ababa, and eastward
from Djibouti and its hinterland through northern Somalia
toward the Horn of Africa. All this made it very clear why Asmara
and the auxiliary bases at Dire Dawa and Hargeisa (in Somalia)
were natural choices for front-line defence and counter-attack.
One had only, indeed, to look at the seasonal breeding and
migration maps and compare them with a set of meteorological
satellite photos to see why the swarms move where they do. Thus,
at the foot of the Red Sea and the inner end of the Gulf of Aden
there is often a natural meteorological 'bridge' formed by winds
blowing between Asia and Africa. This is one of the important
invasion routes (not the only one) by which Desert Locusts from
Arabia are able to enter eastern Africa (or, of course, *vice versa*).
Reinforcing the Ethiopian and Somalian locusts, they will then,
under plague conditions and unless controlled, be liable to sweep
westward across sub-Saharan Africa, picking up recruits as they
go. This normally occurs in spring. Another possibility was that
some swarms, if uncontrolled, might mill around in Somalia
and the deserts of Ogaden until mid-September when normally
the north-easterly winds at that season would enable them to
hitch south for anything up to two thousand miles into Kenya,
Tanzania or Uganda. And there they – or, rather, their progeny –

would stay until the following February, when the ITCZ starts moving back to its northern solstitial position.

'Our job,' said John Sayer, DLCO's chief scientist, 'is to clobber the swarms before they can do either of these things.'

At this point, because it illustrates the rapidity with which the plague developed in eastern Africa and the difficulty of 'clobbering' the Desert Locust when nature favours it, particularly in the initial stages of an outbreak, we shall have to go back some months, to the Red Sea coast and a precautionary survey sponsored by UNDP before the plague was known to be imminent. Its leader, Jim Tunstall, was a British locust officer on loan to DLCO. The territory probed was the long belt of quasi-desert stretching southward from the Sudan frontier and bordered on one side by the foothills of the Eritrean escarpment and on the other by the sea. It is one of those many areas of African earth where one wonders how men can possibly gain a living. Yet many do, and it was among their patches of millet and sorghum, standing green and high after October rains, that the Landrover party found the locusts concentrating. Their numbers were at first few but rapidly increased. In one wadi Tunstall and his Ethiopian colleagues began making counts on foot through millet, and on their second day out eight solitarious locusts were flushed in fifteen hundred paces. Five days later over the same distance the number was seventy-two. Their experience in the Wadi Durama was even more striking. On 29 October twenty locusts were flushed in the millet by a similar foot traverse; two days later the count was sixty-seven; four days more and it was eighty. Insects were springing away from the team members at distances varying from two to ten paces. Early in November came tremendous storms flooding the whole coastal hinterland from the Sudan border to Djibouti, where six inches of rain fell in twenty-four hours. Getting around with difficulty, frequently having to dig their vehicles out and sometimes spending a day covering a mile or so, Tunstall's party noted on 9 November that it was no longer so much a matter of trying to count individual insects as of looking for groups; a dozen up to a yard square, and consisting of first- and second-instar hoppers, were found in two traverses. On the next day hoppers of all instars, indicating exceptionally heavy breeding, were seen in four square miles of cultivated land on both sides of the wadi. By 21 November, exactly one month after the team's arrival, the first fledglings appeared in, and on the edges of, the field, where they had gathered into dense groups measuring 400 yards by 200 yards. Tunstall's assessment after three-and-a-half weeks in the field was that some fifty million locusts and hoppers had established grazing rights over an area of approximately 2,000 square miles.

This was a clear case for ground control; unfortunately the survey teams, who had all their work cut out in getting around, were not equipped for spraying until the second week in the month. When a Beaver plane from Asmara sprayed fifty gallons of dieldrin in its first day in offensive action on 15 November, the kill was minimal. More heavy rain had put the only available air-strip at Gulbub out of service when a reconnaissance by the Beaver was attempted at the end of November: all the pilot saw was a vast sea of green vegetation and a good deal of standing water impeding Tunstall and Co. With the arrival of an exhaust sprayer for attachment to the Landrover and a supply of BHC pesticide for hand-dusting, the ground survey merged itself into a control operation.

The vehicle exhaust sprayer is a device of Sayer's whose use for ultra-low volume spraying has travelled round the world. With a quarter of a pint it can blanket an acre. By the middle of December 277 groups of middle- and late-instar hoppers and fledglings had been eliminated either in this way by dieldrin or by dusting with BHC. Yet so heavy was the foliage screening the growing multitudes of insects that when a third air reconnaissance was flown on 8 December only one small dense band of yellow hoppers was sighted; also a flying swarmlet. On the following day a fourth survey flight dipped low over a plain of heliotropium near Gulbub, where a lot of locusts were known to be present, but apart from flustering several camels which in their turn flushed numbers of locusts as they cantered off, again nothing of significance was seen. Meanwhile, on 5 December, the ground team found their first swarm, a dense one of mature and maturing locusts settled over a square mile at the eastern end of Wadi Durama. Within a couple of days parts of this swarm were laying at the rate of fourteen pods (1,400 eggs) to the square foot. By mid-December nine swarms had been seen and eggpods were being laid in one case at the rate of forty-nine to the square foot.

By this time the airmen had got the measure of the situation and their later spraying sorties were highly successful. Nevertheless, as a means of survey during a key period of the outbreak, the airplane had been found wanting, and it is difficult to avoid concluding that in all such periods, when rain and lush vegetation suggest the presence of locusts, the closest possible integration of ground survey, and ground and air control, are essential. This system was adopted promptly by DLCO after the true state of affairs had been discovered and had it not been for the influx of locusts from other quarters the flare-up might have been arrested. The intensity of the efforts is told by some figures: Between early November and mid-March, when the ground parties – now including three from the Ethiopian Ministry of Agriculture – were working every waking hour in their attempt to get control of the situation, their vehicles covered over 94,000 miles, used nearly 9,000 gallons of fuel and discharged 2,060 gallons of liquid insecticides. Labourers and peasants spread 9,240 lbs of BHC dust. In addition to five exhaust sprayers working out of their desert camps, a DLCO mobile unit used two spraying Landrovers. The Beaver and two other aircraft, Piper Cubs supplied by the Ministry, flew 91 sorties. A total of 4,393 hopper bands of various sizes was destroyed. Barrier spraying, in which swathes of pesticide are laid down at intervals of a kilometre in the hope of killing bands of hoppers advancing into the poisoned vegetation, was also tried. There is some dispute about its success. No doubt many hoppers were eliminated but many also survived to join the mature and maturing adults arriving thick and fast from across the Red Sea and the Gulf of Aden. In January sixteen copulating and egg-laying swarmlets were reported and attacked and hopper groups were being found by the thousand over four hundred square miles. In north-west Somalia, also, breeding was continuous and repeated sorties were being flown from Hargeisa, Dire Dawa and Asmara as the plague expanded. A lull in March raised hopes that the worst was over. Instead, the combination of breeding by escapes and a fresh influx, probably from a district north of Aden which the control workers there had not been able to reach, caused a still more severe flare-up. In April and May swarms were drifting up against the foothills of the Ethiopian escarpment. There and in the highlands fifty-three swarms were reported in May alone. By day they were thinning themselves out into vast veils which could be up to fifty miles long and ten miles wide. Toward evening, when they concentrated and became good spraying targets, they thickly coated trees, shrubs and ground plants over a total area of some 390 square miles.

I flew over this area soon after my arrival, when a report had been received of a swarm being seen on the edge of the Danakil desert near Lake Giullietta about an hour's flight to the south-east of Asmara. The plane used for reconnaissances of this kind was an elderly two-engined six-seater Aero Commander, 'Poppa Yankee', flown by Wordofa Abebe, DLCO's chief pilot. John Sayer navigated. Leaving locusts aside for the moment, it is difficult to imagine another short flight anywhere in the world showing such extremes of scenery. Asmara stands so airily close to the edge of a vast plateau that one is only in the air for a minute or two before the checkerboard of cultivated fields is broken off sharply by the plunging crags of the escarpment. It is not everywhere a clean break for beyond the first ravine there is a whole archipelago of isolated plateau fragments, virtually islands encircled by cliffs of such dizzy height that it is difficult to see how the inhabitants, whose fields spread right up to the edge, get in or out. Probably many, being subsistence farmers, live all their lives in isolation. These were the people the locust threatened. Further to the south millions of years of erosion have crumbled the country into a maze of stony hills nicked with rare cultivation terraces cut into slopes far from any habitation and intersected by ravines into which no roads penetrate. Further south again, the plane windows look one way over miles of dismal steppe and desert to the Red Sea and on the other toward a beautiful medley of crags and mountain passes through which the main road crawls toward Addis Ababa. We turned in over these, flying at a height of 11,000 feet between and over walls of rock that seemed to press upwards and inwards as we skimmed up from one valley then heeled, wings at forty-five degrees, to avoid more rock

15*a* Modern methods of control call for air and ground spraying using highly concentrated poisons such as dieldrin or BHC. The technique of ultra-low-volume spraying, originated in Ethiopia, enables the pilot to achieve a very high kill of locusts with a very small quantity of liquid pesticide drifting through the swarm on to the foliage where the insects will alight. Hopper bands can be attacked by using spraygear working through the exhaust of a Landrover.

15*b* For close-quarters work on smaller infestations BHC dust can be distributed by handblower. This is a cheaper but less effective process.

16 Baiting by BHC dust mixed with bran or groundnut husks is useful against small infestations of hoppers, particularly in districts difficult of access. (*a*) Loading up in Eritrea for a camel sortie into the northern highlands. (*b*) Hand-spreading poison dust among young hopper bands near Shendi on the Nile.

walls before emerging from the pass to plunge into the valley beyond.

The name of this place was Mai Ceu. The DLCO air team, consisting of Abebe and a young Dane, Jan Rahbek, with Sayer in charge of the operation, had, about ten days earlier, located a swarm of twelve square miles settling in two wadis meeting more or less at a right angle between 12,000-foot peaks. This is a typical situation and it calls for a hair-raising spraying technique. Because there is rarely room enough for a long spraying run in the wadi the pilot must gain his speed by diving down the mountain face, clearing its rocks by fifty feet or less, and then flatten out into the upper layers of the swarm, which when settling may be only fifty to sixty feet in height. During the three days Abebe and Rahbek were battling with it, the fight shifted from wadi to wadi, for the escarpment contains myriads of spurs around and between each of which the locusts rest and feed. They are, moreover, moving day by day. Each day, therefore, presented the pilots with a different set of problems, all centering on the cardinal one of how to get at the swarm in this cut-up, tangled and twisted terrain without committing the split-second error that could mean disaster. It is a type of spraying which had never, in fact, been attempted before the beginning of the mass invasion earlier in the year, and has been done nowhere else in the world.

I was later to watch these two pilots working together – Abebe solid and calm, Rahbek wiry and lively – and see how well they complemented each other. As they took turns to roar down the mountainside, each, when flattening into the upper layer of the swarm, would press his stopwatch, at the same time turning on the spray and glancing at the flowmeter as it began to measure the diluted dieldrin flowing from a 120-gallon tank just behind his head. Below and behind the plane the poison, atomized by the spinning cages of the spray units, would begin drifting in a fine cloud down through the swarm on to the vegetation where the locusts were settling. The stopwatch would tell the pilot the time he could safely spend on a given run (it might be only a matter of forty or fifty seconds); the airspeed would indicate its length, and the number of runs would later enable Sayer to calculate the area of swarm sprayed. Crosswise spraying is nearly always preferred

M

because most afternoons at this time of year the wind blows
from the direction of the sea and therefore drifts the droplets
up the valley, ensuring the maximum of kill. Obviously this
increases the difficulty and danger, for most of these wadis are
narrow and deep. The pilot must be sure, therefore, of his escape
route before he dives, and at the end of the run he will have only
seconds in which to climb out through his chosen gap in the crags
on the other side. Another great hazard is the spatter of locust
bodies on the windscreen. Viewers who saw Izaaks' fine film will
remember the machinegun-like succession of sharp-soft explosions
as the BBC sound engineer recorded the noise of the rain of
locusts filling and yellowing the screen until it was almost blocked.
Scores of locusts may also clog up the cooling fins of his engine,
causing it to heat up. All these are factors he must consider while
turning and preparing for the next run about a kilometre parallel
to the first. If, as usual, the spotter plane has remained overhead,
he will be talking to it by radio, as necessary.

Peering through the windows of 'Poppa Yankee', it was easy to
see that one of the great difficulties of operations in this wild and
broken country is the supplying of fuel and pesticides in sufficient
quantity to enable the pilots to keep going while the swarm is more
or less in the same locality. The maximum practical range of the
planes in use is one hundred miles with a load on. Mai Ceu is
more than 200 miles from the supply stores at Asmara. Sayer
therefore had to bring up his replenishments by lorry over a somewhat alarming mountain road to Makale, the small provincial
capital of Tigre province, where there is a reasonably good air-
strip. Although the Beavers have a rather high landing speed for
this sort of country, Abebe and Rahbek were able to fly sorties,
killing perhaps a hundred million locusts. Scanning the scrub as
we skimmed it, Sayer and Abebe concluded that the destruction of
the swarm might have been as high as ninety-five per cent. Abebe
said, 'We certainly hit them hard!' But it is difficult to see
scattered locusts when flying at a hundred miles an hour. Sayer
thought enough might have got away to reform as a swarmlet and
that these could be the locusts reported from the Danakil.

We planed into the airstrip at Makale hoping to get further
information from the provincial agricultural minister whose mule-

eer scouts had brought in the report, and were invited to lunch at
he palace.

H.H. Ras Mengesh Syum, governor-general of the province of
Tigre, is a descendant of a former ruling Ethiopian dynasty and is
married to Princess Ida Desta, grand-daughter of Emperor Haile
Selassie. A small, neat man of quick and genial wit, he has the
reputation of being a highly conscientious administrator. We
lunched on the terrace of a modern wing built onto the old
square brown-stone palace looking out on a green hollow in the
stony bare hills over which we had just been flying. Below the
terrace wall a pair of caged Ethiopian lions, tokens of his royal
lineage, offered occasional roars of homage. In the hollow a group
of village people were building a big round haystack in front of a
grove of eucalyptus trees. I found myself thinking, as so often in
Africa, that such idyllic scenes must be almost the precise counterpart of English country life a couple of centuries ago before the
Industrial Revolution destroyed forever the image of the patriarchal squire and his tenant. But they are, of course, no more truly
idyllic that they were in Cobbett's young days.

The Ethiopian equivalent of the sturdy English peasant is a
poor farmer entirely depending on his land for subsistence, with
perhaps a little surplus to buy the occasional finery for his wife in
good years and none in bad. His average annual family income is
about $480 (or £192). He plants on the rains beginning in May
or June and reaps his harvest in September or October. Ras
Syum recalled that in 1958 the locusts had reduced the whole
country population of Eritrea and Tigre to beggary and famine.

'We are always in the same quandary,' he said; 'we pray for the
rains, rejoice in them and at the same time fear the locusts. Every
farmer, whenever a plague has begun, goes about with terror in
his heart that he will one day find everything gone – his barley,
maize and sorghum, his pawpaws, his green crops, all of them
eaten to the ground or damaged beyond recovery. Yet we must
have the rains. All that we can hope for is that the locusts will miss
us or, better, that the locust control people will destroy them before
they can reach us.'

He had been out the previous day talking to farmers who com-
plained that the ground where Abebe and Rahbek had been

spraying the big swarm smelled so badly that even their goat
would not go on it. 'Which would you rather have,' he asked, 'the
smell of dead locusts or locusts alive in your fields?'

From the airstrip we headed eastward, flying low and vainly
searching the ground for any residue of the Mai Ceu swarm.
Approaching Lake Giullietta – a hazy expanse of salt rather than
of water – the sparse vegetation ceased entirely. We flew along the
rim of the Danakil Depression, which is one desert I have no desire
to see at closer quarters. Blackened by the outpourings of a live vol-
cano on the far edge, its sinister salt surface sent up waves of heat
such as I have rarely experienced elsewhere. For a few minutes
Abebe took 'Poppa Yankee' down to a *minus* altitude – 100 feet
above the ground but 289 feet below sea-level. Then we rose
again toward the welcome coolness and greenness of the escarpment and received our first practical lesson in the significance of
the Inter-Tropical Convergence Zone, which since we left Asmara
had moved into a position where its front of dark black clouds
completely concealed the crags behind which lay the airport. This
was a bread-and-butter matter to Abebe, who now turned out
again toward the Red Sea and then began circling the front until,
finding a break, he was able to fly into Asmara behind it.

Blue eyes gleaming with meteorological zeal, Sayer pointed out
how the front, being easily visible, formed a guide-line for pilots
searching for swarms settling under it. Over parts of eastern
Africa, he said, the width of the space between the two opposing
winds can be as little as three miles, a phenomenon astonishing to
pilots unfamiliar with it, who find themselves flying in a south-
westerly at one moment and a north-easterly ninety seconds later.
Its importance in the life of locusts is that it has a gathering and
stacking action on any swarms around. Taking off in the morning
after a night's roost on the Somali plains, for instance, they may
first travel with a south-wester, streaming along with it in a thin
formation often very difficult to see from the air. About mid-day
they are swept up into the convergence zone and, as the winds
rise, up they go too, packed between the two faces of the weather
front until the topmost locusts in the swarm, unable to fly any
longer because of the drop in temperature, are forced to glide
down to the bottom, when, flying once more, they are again fed

into the upstream. This may be repeated several times over, the locusts below constantly replacing those at the top, all going more or less through the same movements until what was originally a thin, elongated low-flying swarm has become a high-flying and compact one. Later, as the day cools down, the locusts again begin to settle and it is at this time, while they still retain close cohesion, that they can be most easily sighted and, if not too far from base, attacked.

The problems of transportation caused Sayer, even before the last plague ended, to put his mind with redoubled effort to the subject of ultra-low-volume spraying, in which he is an outstanding expert. Obviously, the greater the concentration of pesticide in a given dilution the easier will be the haulage problems. The question of the oils used as carriers in the dilutions is also highly important, for if they volatilize too readily the effect of the poison will be lost. In his laboratory at Asmara, therefore, Sayer aimed to find formulations containing the highest possible concentration of pesticides with the maximum involatility.

The pesticides used in the eastern African campaign, and indeed almost everywhere else, were dieldrin and gamma-BHC, the former being about three times as toxic as the latter but both having the quality of persistency which enables them to keep on killing locusts for days after spraying (they may also kill mammals, but the ultra-low-volume technique of spraying is reckoned to be a safeguard). All the formulations used by DLCO were made to Sayer's specifications. He also adapted the Micronair atomizers to suit them. At Mai Ceu the pesticide used was twenty per cent dieldrin in a solution of fine oil and when the pilots took off they were carrying enough spray in one load for a theoretical kill of a hundred million locusts. This would not happen all at once. The initial spraying would probably kill the locusts at the rate of about 200,000 per gallon. A high proportion of those in the line of attack would receive sub-lethal doses, however, with the result that succeeding sprays would be more effective. The second attack, therefore, might kill as many as a million per gallon. A gravid female receiving a sub-lethal dose has been found, moreover, to pass on enough dieldrin through her system to kill her hatchlings. It should also be possible, Sayer thought, to kill a locust swarm by

spraying into the bottom levels of rising air when the locusts were
caught between converging winds. The poison droplets would then
be carried up with them and as the locusts at the top of the air
stream glided down to be fed into the bottom again they would
receive their second dosage without delay. The plague ended
before he could thoroughly test this method.

The implications of Sayer's 'aerialist' argument, that the locusts
should be allowed to form themselves into king-sized swarms instead of being picked off in penny numbers by the baiting, dusting
and ground-spraying of hopper bands, is appealing in that the
bigger the target the easier it should be to find and destroy, and
as the rest of the campaign was to show, these pioneering methods
and the daring flying of Abebe & Co. were very successful.
Nevertheless, it is not all that easy in practice to find even a large
swarm which may be moving any distance between six and eighty
miles a day on a weather front a thousand miles long. Moreover
it does not need a big swarm to ruin a village community. A
hundredth part of a small one can do so in a few hours. Nor need
it be a swarm at all. A band of hoppers in a field of young millet
can strip it almost in no time, and when this begins to happen it is
no consolation to its owner to be told a spray plane is operating
fifty miles away. On the contrary, he will be far happier to see
the arrival of an old-fashioned control team carrying bags of bait
on camels. The airplane may be more effective. But the camels
or better still a Landrover with spraygear, is a visible token that
someone is doing something *here and now*. This is a factor which
no government of a country of primitive agriculturalists can ignore
and it was not ignored by DLCO's Ethiopian director.

Chapter 12

The Campaign in Eastern Africa:
2 – Toward Sudan

Meantime another problem had been bothering Anthony Izaaks and his television crew for some days after their arrival. Surrounded by locusts and rumours of locusts they had not, as yet, been able to get inside a settling swarm. Yet this, for a cameraman, is the only way to present the reality of the plague. The difficulties were obvious the more we saw of the country. One might, and did, chase a swarm only to come up with the tail of it just as it disappeared into totally impenetrable scrub. We might, and did, see from the air the thin haze of a scattered swarm, wings scintillating, and this would be quite sufficient to enable Abebe or Rahbek to go on a spraying trip. The BBC's Ian Stone and my own photographic companion, Gianni Tortoli, had somersaulted all over the hills and vales photographing such operations from the air. It seemed, at this season, as if there was only one way to be sure of locating locusts on the ground and that was to catch them before they flew. Satellite photos of the weather front passed daily to John Sayer by the meteorological men at the U.S. Air Force tracking station showed the path being taken by swarms; it should therefore also indicate where we could hope to find their progeny.

On 21 July we set out to look for them in the northern highlands. A small convoy of Landrovers and trucks carried food and camping gear for a week. This is 'bandit' country, but in contrast to our expectations (and, I suspect, to Tony's hopes, for he was now athirst for adventure), the trip turned out to be safe and comfortable. A very good road with fine and romantic views across the valley of the River Anseba leads northward from Asmara for some sixty miles to the little city of Keren, where we had arranged to meet Jim Tunstall: blue-eyed, fairheaded, with an air of having walked out of a Graham Greene novel, he was at present in charge of ground survey and control in the northern highlands, where he had been living in camp. Like so many African towns, Keren is a rickety mixture of old and new. Above the market place a piece of rough ground serves its youngsters for strolling and offers them the only horizon most of them will have in their lives – a view of blue mountains, very beautiful if you don't have to regard them as a prison wall. A group of boys about fifteen years old tacked themselves on to me while the photographers were at work. One, with an eye damaged by trachoma, said he had gained his school certificate, adding 'But what now? What is there for me to do here?' His good eye lit on a sliver of yellow among the stones and he pounced, straightening up with a mature locust cupped in his hands. He said a swarm had passed through the district two days before. It was perhaps one of the swarms we had missed on the plain.

I was arrested in Keren for no very clear reason, except that I was cautioned to photograph only locusts, but soon set free. A more serious matter, in this politically unsettled province, is the strict curfew. This is imposed from 6 p.m. to 6 a.m. and at Keren includes a house curfew – lifted for us. For the locust control teams it had the consequence that, since few of the local people could now go far from their villages, an important source of information about swarm and hopper band sightings and breeding was partially closed. Tunstall, in consequence, was relying more and more on his camel scouts, particularly in the north, where the broad shallow waters of the River Barca surged in a brown flood through about 10,000 miles of country completely devoid of roads. A report having reached him of early-instar hoppers being seen some five days' march to the north, he decided to send off a scouting party. Next day, just beyond Agordat, a sprawling dusty town memorable only for the sunlit tower of its mosque poked white-hot into a thunderous sky, and a pretty little girl cadging pennies at the inn, we therefore turned off to an encampment in a palm grove by the river.

The normal muster of a camp such as this, said Tunstall, would be eight camels and scouts plus twenty or so labourers employed in fetching, filling, carrying and loading bags of BHC bait – fifty lbs per bag. One scouting party was already away. The return of the other was expected hourly. They arrived, splashing across the ford while we lunched. Jim interrogated them while they loosed their camels to graze and they then prepared to turn about. Ordinarily a party like this would quarter the country for a fortnight, going from village to village, questioning the people and searching the bush for hopper bands, which they would then attempt to destroy by strewing the poisoned bait. When the hoppers are nearing the fledgling stage this is a hit-and-miss business for they then eat little. Tunstall therefore urged all haste on Mahamed Shink-ahai, the veteran team leader, asking him to go direct to the reported infested spot: 'Otherwise I am afraid you will be too late.'

The bags were slung three aside of each animal. Up went the riders, up behind them went blankets and rations and up off their knees and into the water again went the little convoy of camels. Up on two other camels, less dexterously but very creditably since they had to ride without handholds, went Ian Stone and Gianni Tortoli, taking it in turns to photograph the scene. On the shore, pointing his directional microphone, shaped like a truncheon, toward the diminishing noise of sloshing hoofs, Eric Treasure, the BBC sound engineer, listened with darkling face to the sounds coming through his headphones. It was the beginning of another little locust war. 'All I'm getting,' he said, glaring sombrely at Gianni coming ashore, 'is the clicks of that bloody Nikon.' We drove on in silence, tunnelling through successive walls of rain to Barentu, a district administrative centre with a pleasant arcaded square where a bar served rum, and then on to Gogni before the barrier closed. There, in a disused schoolroom, a charming, round-faced, gentle Ethiopian Muslim beamed at us through his spectacles, welcoming us to his headquarters. Tunstall introduced him as Hadji Hamed Said, of the Agricultural Department, Locust Control Section, Keren. He told Tunstall that he and his men had found a heavy concentration of late-instar hoppers, some of them

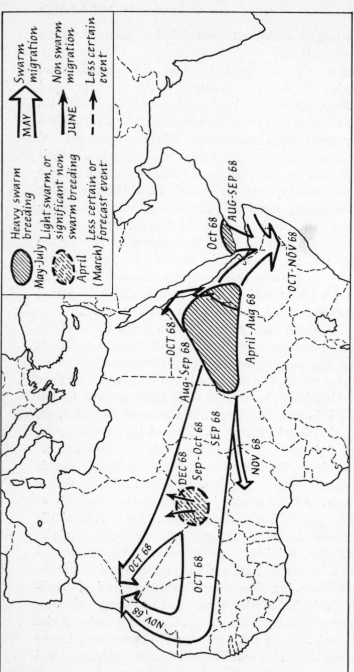

8 *a* and *b* Migrations and Breeding, August/September 1968–September 1969. The collapse of the 1967-8 outbreaks began in autumn 1968 and was virtually complete by the spring of 1969. In spite of stringent control there were some escapes and a weak migration of swarms originating in Sudan and Eastern Africa reached West and North-West Africa, where they were dispersed in the foothills of the Atlas Mountains. Residual populations of Desert Locusts, reverting to the solitarious phase, remained in isolated Saharan wadis, where there were again signs of a slight upsurge.

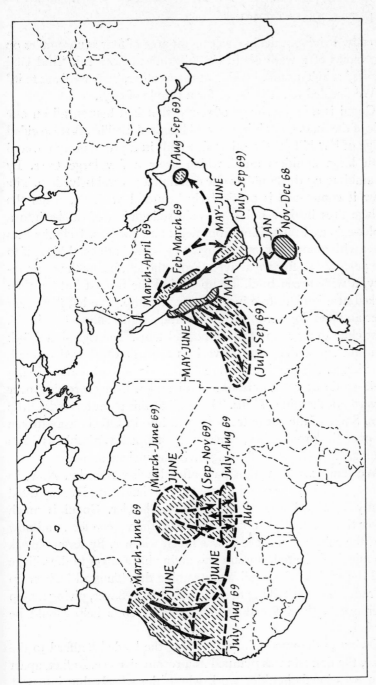

already fledging, thickly covering an area of four square miles on the banks of a wadi about six miles north-eastward. 'And God knows,' said Tunstall, looking at the weather, 'how we'll get to it.'

We decided to settle down for a couple of days.

Gogni is a large village of round thatched houses, all on one side of the road, containing several hundred families. A steep-sided ridge of low hill on the other side ends in a promontory covered with large boulders amid which grow a few large trees. By scrambling up the boulders one can find a natural belvedere. The view it commands is particularly beautiful at dawn, when the village fires begin sending up pale stalagmites of smoke into a violet-coloured mist rising from fields and gardens. I climbed up there with a small, sweet-smiling boy who had attached himself to the party soon after we arrived. From the left, below the promontory, a wide green track led up behind the nearest houses, then forked at a baobab tree just before our camp. The left fork went over an open slope cropped by a solitary camel silhouetted in the mist close to a group of black nomad tents. The right-hand fork continued past the camp to various open spaces dotted through the village. Men and boys were on the stir on both paths driving their cattle down toward the wadi and a bridge crossing over toward another line of hills. The wadi at this point formed a little pass. Some of the cattle trotted down into it. Others were driven up over the promontory, where they stepped as nimble as goats among the boulders, following invisible ancient paths.

We had crossed a high pass after leaving Agordat. Now, in spite of the hills, one could see that all this landscape is really tilted gently westward toward the Sudanese border. Gogni is much lower than Asmara, probably not more than 3,000 feet, and now that the rain had stopped it was much warmer. By seven o'clock the mist had cleared, the sun was giving hard edges to the village and the surface of the soil was steaming dry. Tunstall had set up his radio and was talking to Adefris and John Sayer, giving a map reference of the infested wadi in case Abebe's help should be needed.

Owing to the uncertainty of getting the loaded Bedford to the scene, the first trip was planned as a reconnaissance. In fact, apart from occasional shoulder work where the wheels dug into wet hollows, we managed it fairly easily. The infested wadi, a branch of the Magrab, wound its way through grassy woodlands about a mile beyond a small village called Leda, noisy with a colony of Abdim storks. Hadji said it was the people of Leda who had given him the first information about the hoppers, then in their first instar, a month before. We would have found them now simply by watching the storks. From a tree in the village centre they were flying a commuter service to and from the wadi, gorging on the pink bodies clustered like animate fruit on every spray of thickets of wild palm. By daily hand-dusting and spraying by Landrover, Hadji's team had managed to narrow down the area considerably, but one had only to look at the survivors, teeming, pullulating, scrambling one across another, to see the limitations of ground control when used alone. All the insects were in their fifth instar and many were fledging. Like acrobats divesting themselves of garments in a circus tent roof, they clung precariously to twigs, bodies suspended, struggling out of their fifth-stage cuticles, arching back to the twig again, opening out their wings, drying them in the sun, closing them again, settling in masses ten to a twig, a hundred to a spray, thousands to a palm clump and so on through every clump along the wadi until they added up to millions. Everywhere we looked the same process continued. We had arrived at the precise time when a concentration of *Schistocerca* hoppers becomes a swarm preparing to fly. But they would not do this just yet. Alem Seghid, DLCO's supervisor of scouts, who had been driving me, scooped up quantities of hoppers in a net and, examining the individuals in the catch, estimated that fledging would continue for three or four days. He said: 'The fledglings will go on behaving like hoppers for the next twenty-four hours, then they will begin to fly about a little, but only for short distances.'

We returned to Gogni. Discussing the matter over the camp fire, Hadji expressed the opinion that the hoppers were the progeny of a flying swarm which had first been reported seven weeks earlier, on 3 June, crossing from east to west near Gogni village. He thought the swarm could have arrived from anywhere outside the province, but Jim Tunstall was fairly sure it was an escape from control operations near the Red Sea coast. Then, for a time, we spoke no more about the plague. A villager having wandered up

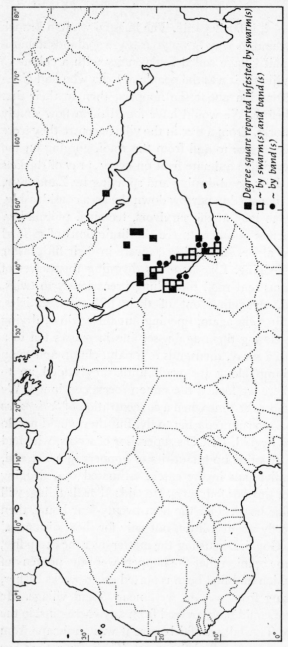

9 *a*, *b*, *c* and *d* Distribution of Swarms and Hopper Bands in a Plague Year, January-December 1968.

As in former outbreaks, the Red Sea coasts early became the epicentre of the 1968 plague, which rapidly spread along the coastal plains of the Gulf of Aden and Arabian Sea and then into the highlands of Ethiopia. Central Sudan also became heavily infested. Swarms which probably originated in Southern Arabia crossed the Gulf to Iran, Pakistan and India and then died out. The West African swarms were almost certainly migrants from the east. They were heavily attacked *en route* and only small swarms reached Morocco. (Above, January-March; below, April-June. See also pp. 178, 179.)

Legend:
Degree square reported infested by swarm(s)
■ ~ by swarm(s) and band(s)
□ ~ by band(s)
● ~ by band(s)

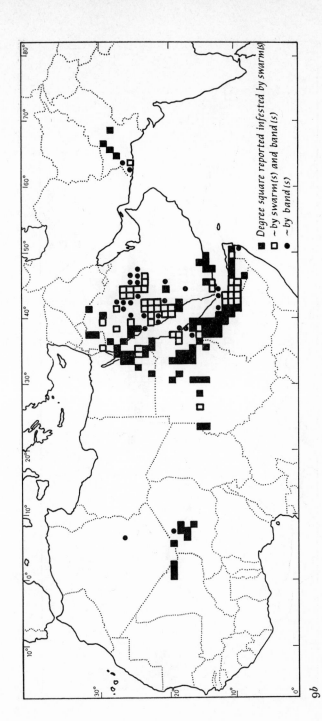

Degree square reported infested by swarm(s)
■ ~ by swarm(s) and band(s)
□ ~ by band(s)
● ~ by band(s)

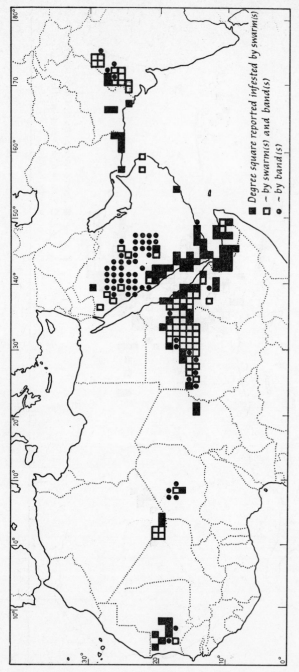

July-September, 1968

Degree square reported infested by swarm(s)
~ by swarm(s) and band(s)
~ by band(s)

9c

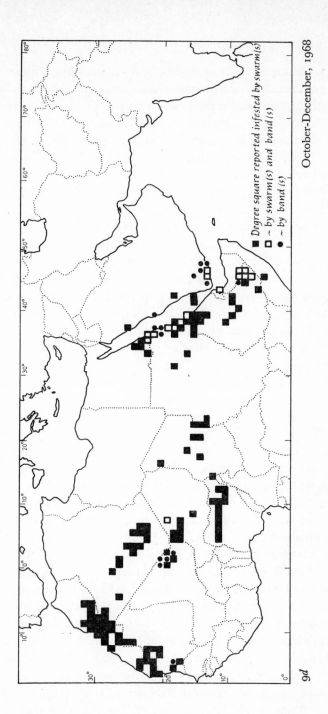

N

October-December, 1968

Degree square reported infested by swarm(s)
~ by swarm(s) and band(s)
~ by band(s)

9d

with a masanco, a five-stringed musical instrument rather like a lyre, we listened to its plaintive twanging while he sang softly of the difficulty of being a poor man in a far-off country where sometimes locusts were all he had to eat. He invited us to a gathering at his house the next day, when he would sing more tunes. On behalf of the BBC, Tony accepted. By first light next morning we were again back at Leda.

Beyond the hopper-infested wadi rose a line of rocky hills about 500 feet high and covered with trees and shrubs. Alem Seghid pointed out a large number of Marabou storks wheeling around the crests. Now and again one of these beautiful creatures which have a wingspread of between six and seven feet, would suddenly go into a very fast, steep dive, wings pinioned, then soar up again with whatever it had seized as prey. To Alem it indicated that a large part of the hills, including slopes and top, was also covered with hoppers, quite inaccessible to a Landrover but a good target for an airspray if the wind was right and Abebe could get in close enough.

At 8 a.m. Tunstall again called up Asmara and, speaking to Sayer, asked for the spraying operation.

'We'll be north, one mile off the Roman Catholic church on the hill. Hoppers are along the bottoms of the hills, about one and a half miles away, as well as on the south side of the wadi. We'll put a Landrover at each end.'

Sayer: 'What about marking them with bonfires?'

Tunstall: 'O.K. We'll try that. What's the weather report?'

Sayer: 'We've got fog and cloud, hope it's lifting. If it's O.K. we'll be along about 11.30.'

At this point Sayer broke off to make inquiries about the aircraft while Jim Tunstall was put through to Tim Wood, Locust Officer at Hargeisa. Wood said there had been an unconfirmed report of a swarm at Las Koreh on the Somali coast more than half-way to the extremity of the Horn of Africa. Locusts in that area are likely to mill around for a long while, forming a kind of breeding reservoir from which other eastern African populations can be topped up. Wood added that he was sending out someone to investigate. Sayer came back on the air confirming that Abebe and his plane would be available, adding that they had sprayed

a swarm previously reported at Batie, near Dessie, on the escarpment 300 miles south of Asmara. It was very thin – 'I reckon it's
had enough now.' He said that at Gulbub, on the coast, the locals
had reported a yellow swarm for which we had searched vainly a
few days earlier. 'We've been up to look but it's pushed off. God
knows where it is now.'

We ourselves pushed off immediately to position the Landrovers
and get the bonfires going. Between our first visit and our second
the fledglings had thickened up considerably and some of them
were now making short leaping flights. Tunstall set out his markers
for the spraying run, fixing the northern end by a dead tree as
well as by a Landrover. This would indicate to Abebe that he
should spray toward the hills, where he would be able to veer
west to clear them at their lowest point. At 10.30 Sayer radioed
that Abebe had left and would arrive at about 11.40: 'That's
pretty near his limit, so he won't stay long.'

The cameramen meantime prepared their positions, Tony and
Ian Stone choosing a spot where fledgling locusts clustered densely
over shrubs round the foot of a tree and coated its trunk for
several feet. 'This will be lovely,' said Ian, 'we'll open on the
locusts, then pan up on the plane as it comes over spraying.'

I then made an important scientific discovery which I doubt
will ever be mentioned in *Nature*. It is that locusts' nerves are
affected in much the same way by the mirror slap of a Nikon
camera as are those of a BBC sound engineer whose microphone
can pick up the noise at a distance of a thousand feet. The locusts,
suddenly aware that Gianni Tortoli had begun clicking at them
within a range of one and a half feet, moved off. The plane was
due within half an hour. This was time enough for me to receive a
deputation, accompany Tony, Ian and Eric in a successful search
for another mass of locusts in line with Abebe's expected flight,
fix up a peace treaty including territorial delimitations with a
clause defining photographic rights (no Nikon to be used while
the microphone was in action and *vice versa*) and return to the
distant sound of Abebe's motor.

Tunstall was speaking to him on the radio: 'Oscar uniform,
Oscar uniform, you are almost there now, Abebe. Fly from the
first fire, it should be just in front of you on the west. . . . You are

completely distorted, I am not reading you at all ... can you hear
me? ... O.K. ... Do two dummy runs. ...' More distorted
conversation from Abebe. Tunstall continued: 'What wind there
is is very light and westerly. You will have to start your run
between the two fires, south to north and then work to the west.'
Ian Stone focused on the locusts, ready to pan up. Eric Treasure
raised his microphone in the requisite direction. Gianni Tortoli,
assured that the noise of the plane would drown his clicking,
capered over a field where a camel hauled a plough. Fifty feet over
our heads, Abebe, having misheard Tunstall's instructions, made
his two dummy runs and then his first spraying run, not from
south to north but from east to west. The BBC got a very good
sound recording of Tunstall's and their own producer's curses and
most of the spray drifted clear.

This was not Abebe's fault. It was the second time we had found
ground-to-air communications unreliable with the radio equipment then available. With some more effort by Tunstall, supple-
mented by his own vast experience, Abebe then sorted out the
situation and made a series of runs in the required directions.

An examination of the swathes of palm thicket sprayed by
Abebe showed that the operation had been effective. Hundreds of
thousands of insects were dying, some needed only another dosing
to finish the work. Half a mile away, however, the case was different, for there in an unsprayed sector, wherever there was a
clearing, every yard of it was pink-carpeted with surging, marching bodies. They were fledglings behaving, as Alem had said they
would, as though they were unaware that, in shedding their last
cuticle, they had ceased to be hoppers. In reality their marching
was more purposive than it seemed, for now and again there
would be a shimmer of wings as one essayed a short flit. Among
those which had climbed on the palm branches these flitting
experiments were much more noticeable. So they would go on,
marching and flitting for two or three days while the final cuticle
hardened; then they would begin making short flights and after a
week would be ready, if allowed, to gather and leave as a swarm.
Thus, the control workers had barely a week in which to eliminate
the threat. Tunstall called up John Sayer in order to report on the
operation and ask for it to be repeated on each of the two succeed-

ing days while Hadji stood by to watch the results. The latter meantime posted his men on the perimeter of the bush. Turning the handles of their dustblowers like hurdy-gurdies, they began advancing through it systematically, being particularly careful to dust the hearts of the palms where Abebe's airspray might not easily penetrate. Little by little the band began to die. We said goodbye to Hadji in the bush and returned to Asmara.

Chapter 13

The Plague Collapses

After our return from the Western Province I was anxious to go on to Sudan, where alarming reports spoke of infestations of swarms and hopper bands worse than any seen for at least ten years. Tony Izaaks decided he would keep the BBC crew in Ethiopia a little longer in the hope of finding more swarm pictures and gathering some local colour – harvest dances, prayers against the locusts and so forth. I reached Khartoum at the end of July and went on by truck next morning to find the situation very gloomy indeed. At the village of Sheick el Amin, twenty miles east of the city, one of the locust officers had the previous day received a scout's report of 'at least five to ten square miles of hoppers' emerging from an egg-field seven miles farther on. We drove on and found that this new outbreak was only part of a generally much larger one. The insects were in instars varying from the first to the third and were being attacked by baiting, dusting and ground spraying. Back in Khartoum the maps showed egg-fields and hopper bands scattered and, in some districts, plastered over a total area of 250,000 square miles. It also looked as though Sudan was setting up an all-time record for swarms and swarm areas. A week earlier Mr Lubani, the Food and Agriculture Organization's regional locust control officer, had cabled to Rome reporting eighty-four sightings of a total swarm area of 767 square miles. Mr Lloyd George, the much-experienced Sudanese government official in charge of national control operations, said one 'moderately dense' pink and yellow swarm sighted in Kassala province had been described as covering an area twenty miles by thirty miles, but this was probably a mid-day sighting when it would have been spread at its greatest. It was very likely the same 'very large' swarm later reported to extend for ninety square miles.

Sudan's troubles (it was thought) had begun with the arrival of six or seven swarms of Desert Locusts from Saudi Arabia in the previous May and June, followed very soon afterwards by an eastward invasion from Chad. But there had, in fact, a year earlier, been a quite serious flare-up of breeding on the coastal plain near the borders of Egypt, where in spite of a vigorous control campaign a fair number of fledglings were believed to have escaped southward. These could have resulted in a couple of generations at a time when the weather conditions were turning so dramatically in the locusts' favour.

Also in June, a sizeable swarm of African Migratory Locusts (*Locusta*) had complicated the picture by unexpectedly appearing at Girba in eastern Sudan, where they ate three thousand acres of sugar cane down to the stalks. Usually this species arrives after migrating across the continent from the Niger. The ravaging swarm at Girba was believed by the Sudanese to have come from Ethiopia, where Jim Tunstall had flushed some scattered solitary specimens and seen a few copulating pairs during his survey of the Red Sea coastal wadis in the previous November. This Girba invasion (if such it was) was particularly unfortunate, for although the sugar canes would grow again (delaying the harvest by perhaps four months), the plane which crashed in spraying them was a write-off. As a result, Sudan's battle was being fought for the time being almost entirely on the ground by fifty trained locust officers and five hundred skilled labourers, in addition to as much local labour as could be hired.

On the day following my visit to Sheick el Amin I flew in a Ministry Cessna 180 to Shendi on the banks of the Nile a hundred miles north-east of Khartoum and half-way to Atbara, at the confluence of the River Atbara. The desert between the rivers is a triangle of 20,000 square miles of practically empty country with its base on the Khartoum-Kassala road approaching the border of Ethiopia's Western Province. As I had noted on the other side,

egg-fields, hopper bands and swarms were being located left, right
and centre of the border. Wherever there had been heavy rain –
and this was practically everywhere – the locusts were laying and
multiplying. Many swarms must have laid in the wadis in the
triangle between the rivers, where the ephemeral vegetation
following the rains made serpents of green almost as intense as the
banks of the Nile.

Shendi is close to the ruins of Meroë, one of the great cities of the
Kushites destroyed by King Ezana of Axum in the fourth century.
The wadi into which we drove eastward was a flat but ethereally
beautiful place, as such corners of the desert often are. Only imperceptibly lower than the harsh expanses on either side, it had over
the centuries caught and held just enough moisture to fill it with a
well-spaced park of shapely, small, thorny acacias. They grew out
of shell-pink sand strewn with translucent green and gold. The
gold (for yellow, even bright yellow, seems far too pedestrian a
description) was a swarm of copulating locusts. Lining the branches
of trees, lying thickly in their shade, coupling on clumps of
perennial grass, on the sands, among the new annual grasses
delicately coating them, in the shadow of the Landrover, *on* the
Landrover, on the heel of Tortoli's boot when he kneeled to take
a close-up photograph, without regard for the nature of their
perch they went about their business. If one made a rush at them
they took to the air for a few yards, still coupled. Or they stayed
where they were. Some of the females had begun egg-laying,
thrusting their ovipositors down into the pink sands where the
needful layer of moisture lay. The males remained on the females'
backs, too exhausted to move or perhaps just basking in the sun.
Had it been possible to call up an airspray at this moment the
swarm could have been wiped out, for it was not very large and
was unlikely to move quickly. All that the locust men accompanying me could do was note its site as an egg-field and hope to send
back a Unimog or a gang of labourers with hand-dusters from the
party working farther up the wadi.

But this party was in no position to divide its forces. They were
working at a place which bears no name on the map but is known
to nomads far and wide as Ain Abujanja – Abujanja's Well. And
there, *at* his well, was Abujanja, in full-throated dispute with a

group of nomads. While their seniors raged, a couple of boys cantered to and fro on donkeys, hauling up water and tipping it into a small canal encircling the well-head, from which it flowed, when holes were unstopped, into a ring of shallow troughs separated by barriers of thorny boughs so that the animals could not jostle and fight while watering. Abujanja had dug and constructed all this with his own hands as his personal gift to the desert nomads and now they were denouncing him, shouting and brandishing guns. Or was he denouncing them? He broke off the altercation and came across to greet us, a grizzled, sturdy man of sixty-two, as strong as a tree, with a tough, lined face and un-perturbed, twinkling eyes. He said: 'I told them to get their animals out of this part of the wadi as soon as they have finished watering them, because I am going to spray it. They say, "Abujanja, you kill the locusts and spoil the grass; we kill you".' He explained that to a nomad a locust visitation is of no great significance, whereas even the poorest grass is all-important. 'I have been tell-ing them that the locust also eats grass. If we do not destroy them the grazing can be gone in a day. Now I must wait to see if they believe me.' They decided, reluctantly, that they did. Shifting their feet uneasily, the nomads, while we talked, had been casting imploring looks at Abujanja. He returned to the group. The leader addressed him: 'Abujanja, you are our friend. We owe all the water, our life, to you. Now, we don't really understand why you want us to go away, but you are wise, we trust you, we will go. But for how long, Abujanja?'

He promised them that in a week, ten days at most, the pesticide would have ceased to be harmful, and that wherever they were he would then send to them. He told them also what parts of the wadi were at present safe and that they could come to the well for water whenever they wished but not to bring their animals. Disappointment – for the grass was so green – was patent in their faces, but the anger was gone. Abujanja had won another little battle in his long career as a locust officer, as I was sure he knew he would. For this was his place – he had been born and named Mohamed Abdulla Mustafa at Tragma, a Nile village near Shendi – and these were his people.

Meanwhile, the insects were coming closer. But they were not the golden locusts. These that came now, like wavelets eddying forward in short runs over a level sandy beach, were a multi-million horde of hoppers, most of them in their first or second instars, about the size of houseflies and not much stronger in colour – the general effect was of a pinkish-grey – and all incessantly advancing. Not all were scurrying with feet to the ground. Over the mass of those who did so the air shimmered to the height of a foot with leaping miniscule bodies, rather like a ground mist of gnats if each gnat were magnified a thousandfold.

Although the civilized mind, and particularly the urban mind out of touch with nature, recoils as by instinct from the sheer mindlessness of the thing, one has to confess that it also has its fascination. This pullulating, leaping and scurrying mass encroaching on the well was an expression of an energy which all through the centuries has been a byword for terror and destruction. It was not hard to believe with Joel that these creatures could by sheer weight of numbers devour the pastures of the wilderness (for this they were visibly doing), desolate the granaries and perplex the cattle (as they had thwarted and perplexed Abujanja's nomads) and bring the peoples to anguish and trembling at an enemy 'like to nothing in all their lives'. Here, close to the Nile, we were witnessing the most historic soil of all being threatened once again by its most historic enemy.

Abujanja's men had paid out a long hose from the Unimog and were using its high-pressure jet of liquid dieldrin rather like a brush, one stroke this way, one that, backward and forward at the head of the mass, then slowly, as its members died, advancing into it, still rhythmically brushing. Where wavelets had escaped (sometimes becoming rivulets) other men finished them off with powered dusters. Here and there in the distance others again were baiting with a mixture of groundnut husks and BHC.

When he was a young man, Abujanja recalled, the only defences against such a plague as this would have been the digging of trenches into which the hoppers would be swept, trampled and burned. Together with poison baiting (sodium arsenate was then used, in spite of its direct danger to animals) it offered protection to local crops but very little more. Everything depended on being able to turn out armies of men, not always easy in desert country.

These new methods, however, were far better than the old ones; but I wondered how effective they could really be against a plague now so obviously running riot – against, for example, the progeny of a swarm Abujanja had witnessed seven days before and which covered, he estimated, twenty-three square miles – or against one he had seen in Kordofan during the previous plague, passing continually over his head for two hours. Piecing together reports from other locust officers and weighing them together with his own experience, he conjectured that the infestation stretched practically continuously for 200 miles from the neighbourhood of Shendi to New Halfa in the middle Atbara valley and covered the whole range of development from egg-field to adult swarm.

We had not far to go to find an egg-field – only half a mile, in fact, on the other side of the well. Abujanja took a spade from a truck and dug into it, lifting a foot of friable soil, very faintly moist at the bottom of the blade, and fingering it apart carefully, counted 43 pods, equivalent to, say, 4,300 eggs, most of which would have hatched successfully. How far did it continue? He shrugged. The situation was clearly getting beyond estimation except by the broadest generalization.

Yet the time factor in control was becoming acute. One of the worries irking the plant protection service in Khartoum was that with autumn approaching much of the equipment now being deployed in the desert against the locust hordes would be needed in the cotton fields of Gezira and elsewhere to protect them against a host of other insect pests. If the campaign against the locusts proved unsuccessful or had to be unduly prolonged, the consequence would be a heavy double burden on the organization and a grave threat to the most important part of the country's economy. For Sudan lives by cotton. It is its chief export, with a yearly value of £50 m.-£60 m., and the most vulnerable time for the crop is just before it flowers, when locust fledglings are at their most voracious. Graven in my own mind, as an illustration of the sensitiveness of the cotton market to the barest rumour of the threat of locusts, is a press message I sent from Khartoum in August 1968 describing the then situation and saying that Sudan's million acres of cotton were 'at stake'. This was misconstrued to imply that wholesale devastation was imminent, and for a day or so the market trembled.

Yet, unrealized by the men in the thick of the battle, the tide was on the turn. The locusts surviving the DLCO and Sudanese summer campaigns would breed again but never, during this outbreak, would they again be a serious menace. But of this we knew nothing at the time. I returned to FAO in Rome in August impressed by what I had seen of control activities but gloomily convinced, nevertheless, that the duration of the plague would have to be measured in years, as all others had been. The meeting of the Desert Locust Control Committee was held two months later in a mood that can best be described as one of apprehension. For the first time in the history of man's age-old battle with the oldest of his insect enemies, however, the pessimists were wrong. The charts prepared later by the Desert Locust Information Service on the basis of reports from all countries in the potential invasion area were to show that the plague reached its peak over a period of three months from June to August 1968 at a time when everything seemed to be in the locusts' favour. Considerable areas of desert were still patched and streaked with green vegetation. Rain was falling in places where further breeding would normally be expected. Yet after a brief explosive increase in the numbers of hopper bands and swarms following the summer breeding their numbers began steadily falling away.

Sudan was the first heavily infested African country to become clear. In the light of hindsight one may say that this was normal, for in past swarm years it has been noted that when summer breeding ends in the central and northern provinces it tends to be followed by a general migration toward the Red Sea coastal hinterland. There the insects fall into the thrall of a complex weather system which may carry some northward to the Egyptian border and others perhaps southward toward the Ethiopian coast. The control teams of the three countries thus have a second chance, and one moreover for which they have time to prepare. In the autumn of 1968 the pattern of these migrations was little different, nor was the position made clearer by a confusion of reporting. Vigorously as the Sudanese campaign was pressed, the size and density of the infestations and the limited means of counter-attack made it obvious that there were bound to be a number of escaping swarms.

A doubtful few may have crossed into Ethiopia from the Tokar delta on the Sudanese coast; more seem to have arrived on the border with Egypt, where the same swarms appear to have been reported several times over; others – how many is uncertain – crossed the Red Sea to stoke up further trouble for the Saudi-Arabian control teams anxiously watching the coastal deserts; while others again, having bred in the west of Sudan, streamed off in echelon 3,000 miles towards Morocco. The result of all these movements was that by the end of September, in spite of fresh breeding, most of the infestations around the Nile and the Atbara and in the westerly province of Darfur were substantially lessening and the threat to the cotton receding. The fact that damage was confined to some 1,500 acres can be counted a success for the control teams, but the clearance of the heart of the country could not have happened without the aid of the winds.

Let us follow the westward-goers, for their flight was a fateful one, illustrating equally their powers of sustained effort and man's new powers of thwarting them.

In spite of the efforts of Abujanja and his fellows, some 800 square miles of young swarms successfully survived the summer control campaigns in central and western Sudan and by late September some were already crossing the border into Sudan's westerly neighbour, the Chad Republic. The Michelin map of Africa marks most of the north of this enormous country with a pink border indicating a prohibited zone. Were a locust capable of reasoning, the swarms could hardly have chosen a better route, for a great deal of this region, including the Tibesti Mountains, the Ennedi massif and the deserts and quasi-deserts far to the south, was at that time virtually a no-man's-land in a civil struggle between the government, aided by French troops, and rebel tribesmen. The Chad headquarters of OCLALAV, the West African locust control organization, were far away at Fort Lamy, the country's capital, and the only practical communications were by radio. FAO, financially aided by the United Nations Development Programme, had a year or two earlier set up a radio network across the whole of the Desert Locust's invasion area and there was one link between the lonely oasis of Fada, in Ennedi, and Fort Lamy, one hundred desert miles to the south-west. It brought the news of swarms reportedly flying on light easterly winds on 25 September. A reconnaissance team was able to get out to the scene later and confirmed that possibly three swarms had passed on their way westward. Their route was governed, of course, neither by reason nor instinct but by the prevailing windflow in the Inter-Tropical Convergence Zone. The next reports, seven days later, came from western Chad, including the long valley called the Bahr el Ghazal, denting the southern Sahara within a hundred miles of Lake Chad. Fortunately for the new plantations along its shores the locusts passed well to the north of the lake.

By 12 October swarms were appearing in southern Algeria, around the Hoggar Mountains. Some may have originated in Niger and Mali; others had almost certainly come from Sudan, now about 1,400 miles behind them. This was a dangerous situation because it meant the immigrants might now be able to pick up local reinforcements. The control organizations of all three countries were ready for the occasion and ground and air control teams were quickly at work. Among the experts on the spot, almost needless to say, was George Popov, who had returned to the same part of Niger, north-west Aïr and the adjacent Tamesna, where he and Roffey had made their classic observations the previous year. He was able to report that by 9 November more than ten thousand acres of hoppers of all instars, and fledglings, were successfully attacked. The advance of the swarms was obviously weakened heavily at this point.

Plotted on the map, their movements look like those of an over-extended army which suddenly finds itself beset on all sides. But unlike an army they could not retreat. Prisoners of the wind, those which survived could only go on. The route ordained for them continued westward and north-westward across Mauretania into the Spanish Sahara and brought them finally to the Atlas Mountains of Morocco. For the progeny of the swarms which had started out from Sudan and others which had joined them this was the end of the journey. As always one feels a certain sympathy with this most tenacious of insects. Collectively they had flown across nearly three thousand miles of deserts and now, in swarms so shrunken that the biggest rarely covered more than a dozen square miles and most very much less, they faced the biggest anti-locust armada that has ever been brought together in one country.

The agricultural riches of the Agadir Province of western Morocco, into which the locusts were being herded on a south-east wind, consist chiefly of splendid groves of orange trees and other citrus fruit. They fill, for example, nearly all the lower lands of the hundred-mile-long Souss valley stretching eastward from Agadir below the heights of the western Atlas. It was towards this valley that the locusts were flying, as they had done at the same season fourteen years earlier, when practically every orchard was devastated and millions of pounds worth of damage done. Once in the valley the locusts cannot get out because at this time of the year the mountain air at 10,000-12,000 feet is too cold for sustained flight. If uncontrolled they would over-winter in the valley and then, in spring, fly northward and eastward in strength to wreak more ruin along the cultivated coastal hinterland from north-west Morocco through Algeria and Tunisia into Libya. There they would breed again and by taking advantage of northerly winds would be able to cross the Sahara from north to south, thus completing the migration circuit, as Professor Pasquier pointed out during our Saharan travels three years earlier. Some of their progeny might then return eastward to Sudan and Ethiopia while others could again go on the ancestral round through Morocco. It was necessary therefore to break the circuit, and the Agadir province was the best place to do it. The Moroccans had assembled five ground teams for baiting and dusting, two groups of three Piper Super Cub (PA18s) spraying planes and one group of three Pawnees (PA24s), to be supported when needed by one group each of Super Cubs and Pawnees and two groups of three Cessna Agwagons – a total of twenty-one planes. In addition the farmers in the endangered areas had been equipped with fifty-four high-speed powder-blowers and 130 tons of pesticide consisting of methylparathion and BHC. The defenders also had a new weapon up their sleeves in the form of a quick knock-down pesticide previously untried against locusts – DDVP (Divorchlos), which is used against flies in butchers' shops. This was sprayed increasingly, and with excellent results, as the campaign developed.

From the beginning it was obvious that FAO's faith in a highly coordinated, international effort, by which widely separated countries were able to launch their counter-attacks in good time, had handsomely paid off. As the swarms came drifting into Morocco from late October onwards the watchers noted that their average size was only about a tenth of those seen in past plague years; their arrival, moreover, was much more gradual. The control teams were never in danger, therefore, of being overwhelmed, but could set about the killing systematically and had time to make a close study of the results of the various pesticides. The campaign lasted approximately ten weeks, and by the end of it swarms totalling 560 square miles had been reduced by at least ninety per cent, at which level most locust experts believe they cease to be viable. Either they break up and disperse into much smaller numbers, with survivors reverting to the solitary phase, or they are destroyed by predators. This is what seems to have happened to a small swarm – one of about half a square mile – which was reported eaten by storks and seagulls four days after the campaign closed.

Elsewhere other battles had also had to be fought.

Locusts which had not caught an east wind out of Sudan were able to migrate, as expected, on a west one to the Red Sea coast. Some met their end at the hands of extermination teams on the Egyptian border while others swarmed across Saudi Arabia and reinforced the local populations scattered through the wild coastal hinterlands north and south of Mecca. This Saudi infestation led to the calling in of the U.S. Air Force on an unaccustomed mission causing some controversy. The machines used were heavy planes of the type used in Vietnam to defoliate forests, and were flown by pilots who, according to Popov, an eye-witness, had little or no experience of the refined techniques of crop or locust spraying. Because the locusts were scattered over a very large area, averaging in some places about one to a hundred square metres when flushed on foot, it was decided to lay down barriers of poison designed to kill any in the direct track of the planes and to intercept and kill others feeding on the contaminated vegetation later. There was also a programme of blanket spraying by which it was hoped to eliminate every locust in certain relatively compact areas. In one of the operations, in the sandy foothills toward the Yemen border, the U.S. planes sprayed 47,500 litres of dieldrin over swathes of country totalling a width of more than a third of a mile for a hundred miles. In all, in this part of the exercise alone, some 207,000 hectares, or well over half a million acres, were treated and probably some forty million locusts destroyed, all in a matter of days. Spectacular as it was – and efficient as it was, for without doubt it played a significant role in ridding this part of Saudi Arabia of the pest – it provoked some afterthoughts: it seemed a costly way of killing locusts and raised anew the question of possible environmental risks involved in laying down such a great quantity of such a persistent pesticide over so large an area. Blanket spraying, whereby the whole of a generally infested area is treated uniformly irrespective of the particular whereabouts of the locusts inside it, was carried out chiefly on the grass-covered sandy plains north of Jeddah and accounted for some 24,500 litres. Ground vehicles using exhaust sprays were on the go elsewhere and by the end of April the Tihama of Saudi Arabia was virtually clear.

In Ethiopia and its eastern neighbours on both sides of the Gulf of Aden there was a long-drawn-out struggle through the winter. Some of the small swarms seen in the northern highlands and Eritrea may have come from Sudan and there was probably an invasion from South-West Arabia into Somalia. The big worry for Adefris Bellehu and John Sayer at Asmara was whether enough of these insects would be able to survive control and start a major migration into Kenya. This has frequently happened in the past; in fact, it is always a possibility whenever the seasonal north-east trade winds arrive on time. The locusts are aided by the fact that between October and January there is usually a period called the Short Rains enabling them to breed successfully. In late 1968 the Short Rains duly fell, the sands of Ogaden again popped with the surviving locusts' progeny and Adefris, Sayer and their aides found themselves confronted with the expected problem of carrying out an extermination programme in one of the remotest and bleakest corners of eastern Africa. Owing to the hostility of the terrain and the uncertain attitudes of some of its inhabitants, most of the reconnaissance and control work had to be done by

o

air, from DLCO bases at Dire Dawa, in the railway area of Ethiopia, and Hargeisa in north-west Somalia. A total of thirty-nine swarm sightings was reported, one of them being said to cover eighty square miles. The locusts were milling around a great deal and some swarms seem to have divided while others amalgamated. If multiple reporting of the same swarm is allowed for, there were probably a round dozen swarms significant enough to keep the DLCO pilots very active indeed during the whole of the critical January period. Altogether they flew their little planes for 172 hours, forty being spent in actual control. Their efforts were very successful. Every swarm located was probably eliminated. Nevertheless, it is believed there were some escapes, for which Nature herself provided the remedy. The usual northerly component in the winds aiding migration into Kenya was unexpectedly absent and the survivors which might have taken that route were most likely carried into the south-eastern mountain valleys and gorges of Ethiopia, where they apparently died off.

By May, in eastern Africa, there were enough blanks among the reconnaissance reports to warrant the conclusion that for all practical purposes this region was also clear. Across the Gulf of Aden, in South Yemen, the previous year had closed with a vigorous ground attack on 400 or 500 square miles of hoppers, after which nothing else of importance was seen. In Pakistan and India, so often troubled by the Desert Locust, the initial invasions of 1968 had come to nothing thanks to a good control campaign followed by a drought. Assuming that the plague had ended and this was not just a lull, the date of collapse can be broadly given as midsummer (meaning the Northern summer), 1969. But was it really the end?

Chapter 14

The Future

The Desert Locust outbreak which began in 1967 lasted approximately two years. This is very short. The subsequent recession, at the moment of writing (December 1970), has continued just over a year; but a perturbingly large number of hoppers, probably running into hundreds of millions of all instars, have been found and sprayed with dieldrin in some parts of the southern Sahara where previous control was apparently successful. This population may have descended from survivors of the earlier campaigns, or they may be progeny of insects which escaped detection elsewhere and moved on.

There have been brief recessions before. A similar spell occurred in 1948. Looking back on it now, it seems to have been little more than a minor interruption of plague conditions which continued, in all, for twenty-three years, from 1940 to 1963. There is, however, a great difference between that recession and the present one. Then, only the crudest methods of control were possible and it is very unlikely that they played any significant part in bringing the plague to a halt. The more likely causes were the failure of seasonal rains and the prevalence of hot, searing winds at a critical period. Of course, in all anti-locust campaigns the control teams must hope for a fair portion of luck of this kind, and the 1969 recession is indeed partly explained by the failure of the north-east trades to offer their usual help in shepherding the survivors of Ogaden down into Kenya and beyond, where control might have been difficult;

likewise the arid months ensuing the early summer campaign in Pakistan and India in 1968, although woeful for the farmers, did at least stop the further spread of the locusts. But by far the greater part of the credit must be given to human agency: sophisticated techniques of control and well-organized international action, including an admirable information and forecasting service – these were the prime causes. For the first time in the history of the Desert Locust, a plague had been checked by human action. But to claim that the problem has been solved for all time would be rash indeed. As we know from instances of other pests which were at one time thought to be under control, Nature has a hearty gift for defending herself. Nowhere is this better shown than in the remarkable ability of growing numbers of insects to resist the widening range of chemical pesticides. That some insects evidently have a built-in resistance has been recognized for more than half a century, but it was only after 1945, when new synthetic insecticides based on wartime gas research began to mist the fields and orchards of the more developed countries, that this unwelcome phenomenon started showing noticeably. This should have been salutary in the sense that it ought to have broken down the blind faith that had attended the introduction of such wonder-workers as DDT and other members of the chlorinated hydrocarbon family, like dieldrin. But nature often works by stealth, and to the majority of farmers at this time it seemed that there was no pest to which one or other of the new compounds could not provide an answer. Unfortunately, when resistance began to appear – one of the first examples was the common housefly – the remedy most commonly adopted was to increase the dose, on the hard-dying principle (not confined to the use of pesticides) that more means better. It meant nothing of the kind. The immediate outcome was the introduction and sale of new and more powerful formulations devised to protect the crop at every stage from seed-growing to harvest. As the treatment has increased, so have the numbers of insects showing immunity to it. By 1966 resistance to one or more groups of pesticides, in one or more areas, had been reported in some 180 species of agricultural pests. By 1968 FAO's working party of experts on the subject were able to count 48 more. Unlike Don Marquis's cockroach, Archie, they no longer needed to lament impending massacre; they were immune to it.

The mechanisms of resistance work in various ways. In broad terms, the fortunate insect is one which is able to interpose a biological barrier preventing the poison from reaching the sensitive organ through which the chemist hopes to kill it. It may be blessed, like some houseflies, with a tegument difficult to penetrate. In other cases the organ on which the pesticide is designed to work becomes, for some reason, decreasingly sensitive. Or the enzymes in an intercepting tissue may be altered in such a way that they detoxify the chemical, robbing it of its power before it can reach the sensitive site. These astonishing gifts of survival are possessed, it should be noted, only by a minority, perhaps a very small one, of the individuals composing any particular species. But it is this minority which counts, for it is this which hands on the qualities of resistance from generation to generation. In effect, the pesticide acts as a sieve, letting the resistant individuals through while killing off the others. Sieve – spray – often enough, and you are likely to finish up with a resistant pest population composing virtually the whole species. This would be bad enough if the resistance of any particular insect were confined to one particular pesticide. But we know now that some individual insects have several different defence mechanisms enabling them to put up a simultaneous multiple resistance to a whole battery of pesticides. Other strains of pests, having been exposed to only one chemical, have been able to develop cross-resistance to others having similar toxological action.

Could any of these events occur in the Desert Locust? There is no present evidence of their doing so, but it would be foolish indeed to assume that they will not. The optimistic view is that because of the nature of the insect, and particularly its mobility, the conditions encouraging resistance will not arise. Unlike most of the commoner agricultural pests it rarely stays long enough in the same place to call for a continuous bombardment. True, a large swarm or hopper-infested area may have to be attacked several times, but as a rule this means that the pilots have found the target too big to be destroyed in one day's attack and must therefore return to deal with it bit by bit. Two spraying runs over the same section of swarm may kill up to ninety-eight per cent of it. The two or more per cent which escape *may* include resistant insects, and these may in due course be killed off by predators or parasites. Either way, it looks as though resistance is not at present being unduly stimulated and one would expect it to be slow in revealing itself.

There is, however, another factor which could theoretically aid the locust in developing resistance: this is known as 'vigour tolerance'. It infers that in some insects the ability to withstand pesticides is the result, partly at least, of natural selection imposed by extremes of environmental conditions. Few insects are better equipped to cope with such extremes than the Desert Locust. Is it possible that some of the members of the species could be, for this reason, more disposed than others towards resistance? Given additional selection by more frequent spraying – as has been proposed – could these become totally resistant and eventually dominate the species? It is a chill thought, for if this were so our chief weapon against the insect would be removed and we should have to find another. It is because of the possibility of locusts developing resistance that the experiment in blanket spraying in Saudi Arabia was viewed with dismay by some biologists who feared it might lead to repetitions, possibly with disastrous consequences.

Another objection to blanket spraying, as every Western agriculturalist is now painfully aware, is that when non-specific chemicals like dieldrin and DDT are used, they knock out not only the target pest but also the parasites and some of the predators which are nature's means of control. (Nature, of course, never envisaged the large-scale monocultures, wheat by the hundred thousand acres, cotton by infinite feddans, rows of lettuces reaching to the horizon, by which pests are encouraged.) In the territories invaded by the Desert Locust, its natural enemies tend at first to be overwhelmed by the sheer weight of numbers; but when a decline sets in they accentuate it. Under favourable conditions they may wipe out vast numbers, particularly in the embryo stage, and this had led to the hope that some form of man-induced biological control might eventually be possible. For example, there are certain bacteria which under the right circumstances cause enormous damage as do fungus diseases. These could doubtless be spread artificially, but unfortunately they require a fairly high degree of humidity to take effect and so would be useless in preventing break-outs, which begin in normally arid regions. Even if rain could be anticipated in any particular part of this vast area, the insect is so mobile that it would in most cases have moved off the treated sites before serious damage was done. (In fact, the possibility of applying the right treatment in the right place at the right time is so remote as to be impracticable.) Similar objections face schemes for breeding up and releasing numbers of Trox beetles and *Stomorhina lunata*, a parasite fly which when present in sufficient numbers is able to destroy a sizeable Desert Locust egg-field.

If the locust would stay put, confining itself to a circumscribed ecosystem, like an island or its equivalent, it might be possible to apply the sterile-male technique of control, in which males sterilized by radiation are released into sexual competition with unsterilized males of the same species. A collapse in the population should follow – if the conditions are right. The method has been applied with some success against the screw worm and the Mediterranean fruit fly and there are some, not as yet very well founded, hopes that it may be useful against the tsetse fly. Against the randomly mating, highly migratory Desert Locust it would seem to be useless. Nevertheless, much work on biological control is now being done at the Anti-Locust Research Centre and elsewhere and new possibilities may turn up. For instance, if fledglings could be induced to mature early, at a time when there was no rain, their eggs would not be able to develop. Being laid on hot dry sand, they would dehydrate. This sometimes happens in nature, but the results occur only by chance. The young locusts could, however, possibly be tricked into it. Rapid maturing is believed to be triggered by a body chemical released when numbers of fledglings come into close contact during grouping. Even if the appropriate chemical could be extracted or synthesized in sufficient quantity, the problems of application in the vast spaces of the deserts would remain. It would also require extremely accurate weather forecasting. However, this may come.

Another idea, not strictly of control or even of biology but closely related to both, is to protect crops from invading locusts by spraying them with a natural repellent like that obtainable from neem seed. This has the advantage that it could be done easily and cheaply by any peasant stocked with a small store of seed and willing to act promptly when locusts are known to be about. The neem is a common tree of India and Pakistan, where it is valued for its shade and also for a curious property of its foliage, which appears to discourage insects. It has been an age-old practice, for this reason, to mix dried neem leaves with grain in storage and with clothes in drawers and cupboards; but, like some other folk-recipes, its scientific basis has only recently been tested. I am indebted to Dr S. Pradhan, the eminent Indian ento-mologist, for details of the experiments he carried out in the 1960s with a colleague, Dr M. G. Jotwani. Both men have much experience in locust work. Dr Jotwani was a Locust Warning Officer and has also been involved, as an entomologist, in insecti-cide testing and seed testing.

In their first trials to see whether the locust would reject neem as other insects appeared to do they prepared two qualities of bitters of neem (crude and pharmaceutical), then sprayed them onto greenstuff. This was offered to Desert and Migratory Locusts, as well as to a variety of other insects, all of which would normally find the untreated food palatable. Each, judged by the amount of the treated food they ate, was repelled in some degree. The ordinary omnivorous locusts were the most discouraged and particularly disliked food sprayed with the crude bitters. From this the experimenters concluded that the factor causing repellency must be concentrated principally in the natural unprocessed seed. Testing this further, they weighed a quantity of kernels into an electric blender, added two hundred parts of distilled water to one part of the crushed kernels and sprayed the solution on more fresh food. This proving even less to the locusts' liking, they finally prepared a number of suspensions ranging from 0·5 per cent concentration of neem to 0·0005 per cent and sprayed them over cabbage and sorghum among other plants. Untreated plants of the same species were used as controls. The Desert Locusts refused the treated food absolutely up to the point where there was only the barest trace of neem (a suspension of one part in twenty thousand parts of water) and then reluctantly ate less than half. The Migratory Locusts were a little less choosy. They began nibbling at ·01 per cent and ate heartily at ·0005 per cent. Both species ate all the untreated leaves.

The opportunity to test the effectiveness of neem against swarms came with the locusts' invasion of Delhi in the summer of 1962. Large areas of crops were sprayed with the ·01 per cent neem seed suspension; the swarms settled on them but failed to eat. Specimens captured from the swarms were then offered food in the laboratory. They refused the sprayed leaves but ate the unsprayed ones voraciously. The collapse of the plague soon afterwards prevented tests in village fields, and the 1968 invasion, stopped by the failure of the monsoon, has provided no further opportunity. Meantime a simple method by which a farmer with a neem tree at hand can prepare his own infusion has been worked out. All he has to do, having gathered and ground his neem seed in advance, is put the prescribed amount in a muslin bag, dip it in a bucket of water and squeeze until the water is brown. He then sprays the crop and, according to Drs Pradhan and Jotwani, should gain up to three weeks' protection. Vast quantities of neem seed, enough to supply every farmer in the endangered areas, are at present going to waste. At critical times for the crop they could well offer a useful form of local insurance, particularly in orchards and nurseries, and would complement the national control efforts.

The outstanding lesson of the last plague was that all methods and plans of protection against locusts depend for their efficiency on adequate early-warnings. The Desert Locust Information Service, operated by the Anti-Locust Research Centre of London in conjunction with FAO, collects its forecasting material from all over the potential outbreak area during lulls and from the whole twelve million square miles of the invasion area during plagues. The fact that all these reports are based on eye-witness observation means there is invariably a certain amount of repetition, leading to uncertainty whether, say, the same swarm is not being reported several times over by different people in different places, or whether, on the other hand, the movements of some swarms or hopper bands are not being missed. The systematic charting analysis and comparison of reports begun by Uvarov and Zena

Waloff over forty years ago enables present-day forecasters to make a reasonable estimate of probabilities, and by combining report data with increased knowledge of wind and weather and their influences it is possible to anticipate invasions in a general way by, in some cases, weeks. All the same, in such enormous stretches of country, where even the largest swarm or band is no more than a needle in a haystack, and a moving needle at that, the best forecasts are bound to lack precision. They can only be as good as the data coming in and because of its inevitably patchy quality a lot of thought is now being given to improved surveying techniques, some of them very advanced. Two which have been tried are air photography, primarily for detecting hoppers and vegetation, and radar for the detection of flying swarms and solitaries. Satellite photography, already useful in providing meteorological information, is being studied for its possibilities of direct observation of suitable laying sites and of swarms in flight.

Aerial surveys of vegetation, pioneered by Roffey and Popov as a means of locating places where build-ups of locusts were likely, have previously stopped short of recording the insects, even when in groups. Flying low at 100 m.p.h., the ordinary fixed or hand-held camera only depicts an indistinct blur, which may or may not be locusts, in between patches of greenery. Yet if detailed pictures could be obtained an immense amount of valuable time would be saved, for it would not then be necessary, as now, to make prolonged ground searches. Instead of a lapse of days before calling up a spray plane, it would often be possible to summon one within hours and the chances of the insects escaping would be minimized. A new method, worked out and tested by the French company Geotechnic, suggests that this will in future be possible. Instead of the camera being fixed or hand-held, it is carried in a sliding cradle designed to 'stop' the picture at a given flying speed. A backward movement of the cradle, coupled with the triggering of the camera shutter, in effect cancels out the forward movement of the plane, so that while the lens is open it remains momentarily stationary relative to the scene it photographs, say, one hundred feet below. It is thus possible, in theory, to get a clear photograph of locusts and hoppers on the suspected ground.

What the camera is being asked to do, in fact, is to supply a detailed picture of (a) the extent and state of any vegetation likely to provide habitats favourable to the locust, (b) the number of insects visible either on the foliage or on the ground between the plants, (c) their colour and state of development as hoppers or adults and (d) indications of their behaviour. Are they, for instance, scattered, dense, patchy, clustered or marching, and in big groups or small? Even in an age of highly sophisticated apparatus this is a lot to demand of low-speed film emulsion exposed during high-speed, low-level movement. Tests were made by Geotechnic at the request of FAO and financed by the United Nations Development Programme. For the first experiments, in and around Paris, a few buckets of Saharan sand were flown in and spread around with dead locusts on them on the laboratory floor. Some hopes had been placed on infra-red film but the locust apparently reflects infra-red rays in the same degree as sand and the choice, for further tests in the open near Fontainebleau, fell on that old favourite of amateur photographers, Ektachrome-X. Used on locusts killed by alcohol injections which temporarily inhibited colour changes, it clearly distinguished the variations of colour in insects of both sexes in various degrees of phase change and maturation. After juggling with lenses of various focal lengths,* one was

* Some technical notes on the equipment used in the trials may be of interest. In the laboratory and French field tests the camera was a Hasselblad 500C, used in fixed positions with 50 mm, 80 mm, 150 mm and 250 mm lenses and positioned to give image scales varying from 1 : 20 to 1 : 300. For the aircraft tests in the Tamesna this was substituted by a French Omera-33 camera operating in a moving cradle designed by Geotechnic. Film used in these tests was Aerial Ektachrome (colour) and Aerial Plus X (black and white); 12 cm rollfilm providing pictures with 60 per cent overlaps permitted stereoscopic examination. Exposures were made at the rate of 5 per second on a roll 20 metres long, processing being done in an air-conditioned room at Agadez.

The aircraft was a single-engined Lockheed AL 60, instructions concerning flying height being furnished to the pilot through a radio altimeter, type Bonzer TRN-70 VME (USA). The photographer checked ground speed by means of a Wild cinematic altimeter. Flying speed was limited to 160 kilometres per hour, with very slight tolerance. Very precise flying was therefore necessary.

Having located the sites of the trials, photos were first taken of the whole habitat at heights providing image scales varying from 1 : 1,000 to 1 : 10,000. Finally, the locust sample photos were made by rectilinear flight runs, each photographed strip being at least 500 metres long and 20 metres wide for the picked which enabled the interpreter to score eighty per cent correct recognition on the most difficult prints.

Armed with this information the whole outfit moved in September 1968 to the Tamesna in Niger, where Jeremy Roffey and George Popov were again teamed up to aid in this novel exercise. Roger Pasquier was also present, together with R. A. Steedman, a locust expert from London. The only uncooperative element was the locusts, which failed to appear in any significant numbers until the operation was over. A quite big and well-populated laying site was then discovered and the teams of technicians and observers thankfully extended their plans to include it. As it happened, this paucity of locusts was not without its negative merit, for although only five hundred to a thousand to the hectare were observed by the ground team the camera did in fact pick out some of them and it is fair to conjecture that the rest were hidden by vegetation. The observer in the plane saw nothing with his naked eye. Even more striking was the test over the gregarious site, estimated by Popov to contain patches of up to 600 hoppers per square metre. Roffey, in the photo plane travelling at a hundred miles an hour, could still see none, whereas the camera was able to photograph groups of densities up to two hundred to the square metre from an altitude of one hundred feet. This was really a very good result, for it showed that when the population of locusts has reached the danger point the camera can locate it.

Much still remains to be done before it can be regarded as a fully efficient field instrument. The camera must be flown while the majority of the insects are basking in the open toward either end of the day, and because it cannot photograph hidden locusts, harbouring in shrubs or clumps of grass, it will be necessary to

---

1 : 200 scale. Flights were made at times between 6.30 a.m. (dawn) and 11.30 a.m., after which the aircraft's shadow was found to interfere with photography.

Preliminary interpretation of the photographs was made on Aerial Ekta-chrome only immediately after processing, using a stereoscope without magnification. Further interpretation was carried out in a laboratory. A score of eighty per cent to ninety per cent of accuracy in recognition was considered satisfactory.

work out co-relationships between those which can be seen and identified and the probable numbers of those concealed. These, moreover, will vary from one species of vegetation to another. The camera's ability to provide identifiable pictures of the natural canopy sheltering locusts will therefore be very important. It can also indicate the nature and state of the soil. In other words, it should be able to offer a quick picture of the ecology of the habitat which will be very useful to the control team, provided the latter includes a well-trained eye. And since the object of the exercise is to enable the control workers to go into action quickly, the processing of the film must also be quick. The ideal would be an adaptation of the Polaroid camera principle enabling the photographer to assess the situation instantly, so that it might even be possible to do the initial spraying at once. This is a possibility for the future. It would call for a degree of competence and skill not always available in some of the developing countries where the threat is greatest. The whole operation, indeed, calls for skill of a high order, not least in the flying of the aeroplane at a constant speed at ultra-low altitudes in temperatures which will often be well in excess of forty degrees Centigrade, with a nasty risk of turbulence. The advantages, however, are too many for this new development to be neglected. Research will have to be extended, methods of application improved and expert personnel trained. It is in such matters as this that the United Nations Development Programme project is so valuable.

The presence of Roffey and Popov in Niger at the photographic trials was also the occasion for the radar experiment. For many years it has been known that radar is capable of picking up large swarms, but until Dr Glen W. Schaefer of the Biophysics Research Unit at Loughborough University published the results of his work on the tracking of single birds nothing more had been done, chiefly because the locusts failed to appear. The new outbreak encouraged more thought about it. Dr Peter Haskell, Director of the Anti-Locust Research Centre, got in touch with Dr Schaefer, who, after some calculations based on the Desert Locust's radar cross-section (1 cm^2), offered the opinion that it should be possible, using a small search radar, to pick up single locusts at distances up to one and a half miles. One of the most remarkable enterprises in the story of anti-locust research was decided on and a joint mission, representing ALRC and Loughborough University, set out across the Sahara. Schaefer and Roffey went the whole way by land in a hard-topped long wheelbase Landrover carrying most of the aerial gear, the display unit and a generator; the rest was air-freighted to Zinder in Niger and carried up by track to In Abangharit, where they all met together on 12 September. Popov was already there, working with the West African anti-locust organization. His job in the experiment was to contribute to the biological information. The whole radar outfit, powered by the generator amidships of the Landrover, was mounted, checked and ready to run within six days. Very soon some surprising results were coming. To begin with, it was clear that Schaefer's optimism was more than justified.

In his work on birds he had discovered that the distortion of the thorax caused by the flapping of the wings produces a peculiar pattern in the echo which, when magnified electronically, is quite distinct from that given off by a bird of another species. This pattern he describes as the bird's 'radar signature'. The question was whether the locust also had a 'signature' enabling it to be distinguished from other aerial creatures and objects, including raindrops. There were few day-flying locusts around, so most of the work had to be done by night. When the apparatus was working by day a number of big blips on the screen indicated soaring birds, such as vultures; smaller, fast-moving blips at night were identified by Dr Schaefer as migrating birds; but in addition there were a lot of smaller echoes, becoming particularly numerous about thirty minutes after sunset, when the solitary, night-flying locust normally begins its flight. The numbers rose to a peak rapidly, then began tailing off, which is also what one would expect of night-flying locusts. Direct identification could, of course, only be made by Aldis lamp or other lights up to a limit of about one hundred yards, but these small bright blips were coming from a considerably greater distance. The analysis of the signatures proved that all the echoes were originating from creatures of the same species and when, later, day-flying locusts turned up, it was confirmed beyond doubt that they were indeed those of solitary Desert Locusts, picked up at distances of up to 4,200 yards.

When day-fliers began to appear, the importance of radar as a detector became even more evident. The greatest distance at which swarms have been seen by the naked eye of an observer in a plane is around seventy-five miles, but this is exceptional, and only possible on a very clear day with a big, dense swarm – and very good eyes on the observer's part. Schaefer's radar picked up an *invisible* swarm, one so thin that it could not have been seen by eye until directly overhead, at twenty-five miles. He and Roffey are now reasonably sure that a high-flying swarm of moderate density could be detected over level country at a distance of at least sixty miles and possibly much farther. Because of atmospheric refraction the radar horizon for an object 3,000 feet above ground is seventy-seven miles. One radar station, operating at this distance, could thus scan over 18,000 square miles with the fair certainty of picking up a major swarm.

On the research side of the experiment the outcome has been to remove some errors and uncertainties while opening the way to yet more questions. The most important discovery has been that whereas day-flying swarms drift downwind irrespective of which way individual locusts within the swarm are flying, the solitary night-fliers appear to orient themselves purposively. Although it was known, on biological grounds, that they finished up in areas of wind convergence, no one could say exactly how they did so. We now know that their flying speed must be added to the wind speed – sometimes with astonishing results. One set of observations made when the wind speed at ground level was nine miles per hour, picked up locusts at 500 to 1,500 feet, where, obviously aided by much stronger wind, they were zipping across the night sky at 33·5 miles per hour. Roffey calculated from the evidence of the screen that on one night locusts at a height of between 600 and 1,000 feet flew 170 miles.

All these, it must be remembered, were individual locusts flying not in swarm formation but alone. A feature of the distribution of their echoes on the indicator was the uniformity of their spacing, especially when there were light winds. But when it turned gusty this appeared to change and the highest concentrations, appearing as diffuse swarms, occurred under these conditions. The inference was that the wind was an important factor in gathering

them together and perhaps, therefore, in their eventual gregarization.

By using radar it should also be possible to arrive at a much
more accurate count of the night-flying solitaries' numbers, for
whereas before only those at the lower altitude levels, within the
range of the Aldis beam, could be counted, it will now be possible
to keep track of those higher up. This also points to the need for
knowing more about the structure and behaviour of winds at
night. Given better information of this kind, it should then be
possible to forecast what the solitaries are doing on an hour-to-
hour, day-to-day basis, instead of, as now, a seasonal one.
Information about night winds in the desert is at present scanty,
but, appropriately enough, the night-flying locust, by acting as a
marker, whose movements show up on the radar screen, can help
to supply it.

Radar will also make it easier to study swarm structure and this
leads to the possibility of more sophisticated spraying techniques
on the lines that John Sayer has envisaged – air-to-air spraying in
which the droplets are carried by the wind in and through all
parts of the swarm. Finally, it has emphasized once again a consideration that was in danger of being lost to sight – that the
Desert Locust, more than most animals, is a creature of dual
environment, terrestrial and meteorological. The more we understand about the elements and how they interact, the nearer we shall
be to achieving full control.

Not entirely unexpectedly, attempts to put the principles of
radar detection into practice elsewhere have sometimes run into
political difficulties. This is regrettable but inevitable particularly
during non-plague periods when the sense of urgency is apt to
cool off. Yet it is precisely at such times that surveys are most
necessary for an efficient early-warning service.

In a relatively crude way meteorological satellites have helped
by indicating major weather upsets likely to produce conditions
suitable for locust breeding and multiplication – the cyclone
spotted breaking up over the coast of Muscat and Oman in
November 1966 is an example. A suitably timed weather satellite
can also, as the locust hunters in eastern Africa have found, give
a very good idea of the weather front where the swarms are most

likely to be. Within the next decade refinements of weather reporting by satellite will have filled in large gaps in our day-to-day
knowledge of wind and clouds in the deserts where present
meteorological knowledge is scanty or insufficiently prompt. This
is basic meteorology, very valuable but still not providing enough
evidence for firm conclusions about what is happening on the
ground below. Using normal photographic methods it would be
possible to get a satellite television picture of some quite small
areas of standing rainwater, but in most cases they would have
disappeared, baked dry by the sun within an hour of forming, and
leaving no visible trace of the vital moisture underneath. What the
experts of the Desert Locust Information Service, plotting their
charts in London, would like would be the ability to 'see' this
hidden moisture, up to say ten centimetres down. They would
then know where rain had fallen and when and where to expect a
subsequent burst of greenery capable of harbouring breeding
populations. With this information in their hands the national or
regional control organizations could start gearing up for spraying
operations and a great deal of tedious and expensive ground surveying would be obviated. The field teams, including pilots, would
know where to go and if the numbers and behaviour of the insects
indicated danger they could be dealt with promptly before
becoming dangerous.

All this is not so remote as it might seem. The instrument that
may give it actuality is the American Earth Resources Satellite.
Designed to probe the earth's surface for such purposes as preliminary large-scale geological surveys, its ability to 'see' into the
soil would equally well serve the anti-locust campaign. Using
special instrumentation, the desired information would be fed
back in such a way as to construct a map of sub-surface wet areas,
even where nothing showed above them. If, in spite of this sophisticated aid, a period of high plague did develop, the same
satellite could provide invaluable information for the forecasting
service by photographing vegetation and possibly, also, swarms.
Knowing where the swarms were, the forecasters would be able
more easily to anticipate their movements.

Yet another possibility that sounds eerily far-off, yet is not beyond
the bounds of reason, is a satellite which would be used to smell out

P

unborn locusts by means of a chemical sensing device. This pre-supposes that eggs laid by locusts in sufficient densities can be shown to give off a characteristic, identifiable gas. It is possible that by means of spectrographs the faint whiff from an egg-field could be sensed by the satellite passing over it. This development may be far ahead. The more immediate advantages of using satellites in the ways described are the saving of time and money in countries in perennial need of technical aid and the ability to track down locusts in areas too difficult to reach or survey. Ideally, surveying by satellite should be a United Nations project. It would thus be taken out of the tricky realm of international suspicion.

Chapter 15

Methods and Morals

In the last week of May 1968 Alem Seghid, superintendent of scouts of the Desert Locust Control Organization of Eastern Africa, was working with a ground team near the River Awash about 150 miles west of Djibouti, when he came across masses of locust hoppers and fledglings infesting the bushes and long grass of a patch of country measuring about twenty-five square miles. The insects were mainly *Locusta* – the African Migratory Locust – with isolated Desert Locusts among them. Alem, whose vehicle was a Landrover fitted with an exhaust sprayer, was able to call up three aircraft belonging to the Ministry of Aircraft and together they gave the area a thorough going-over with a powerful solution of dieldrin. They also dusted and baited. This operation was evidently not fully successful for when they looked at the same spot a month later a two-square-mile swarm of Migratory Locusts was feeding on it, and isolated locusts were scattered all over the area. Continuing the battle for another three days, Alem noted that from first to last they used 1,465 gallons of liquid chemicals and 650 sacks of bait and dust, all on the same twenty-five square miles. Then came a slightly worrying incident, described in his report:

At the last date ... the local people came to ask us on rush – we ask them what happen and the local people said your chemical killed forty-five goats so we went to see at the same area where was being sprayed but I think these people's report was right. The goats were killed by insecticides. But we don't know which insecticide has been killed them because we have five kinds of chemical. . . . I think somewhere places we are being sprayed too much because the *Locusta* they are not moving much they go on circle on the green bush. I think that is why the goat being killed or by feeding grass or drinking water, because that area always swampy. But we went to see Bitweded Ali Mera, the Governor of Dankalia and explain about the goats being killed but he said never mine about this because you were doing the right job – killing the locust much import, never mine killing forty-five goats.

I have quoted Alem's account at length because the sad fact is that he and the Governor were both, in their own ways, right. There *are* too many goats in Ethiopia; it does not matter (except to their owners) that some are killed; and killing locusts is infinitely more important. Alem, with whom I travelled a great deal, is an intelligent and (as his report shows) an honest and conscientious man, as well as being a good field naturalist who was quickly able to identify the cause of the animals' deaths as the over-use of insecticide.* The five kinds of chemicals were probably really only two – dieldrin and BHC – in various formulations, and it is unlikely that any one application would have been lethal. Alem, like everyone else at DLCO-EA at the time, was working under extreme pressure under full plague conditions. The hoppers were evidently at a stage of development when they are reluctant to feed and so were proving exceptionally difficult to kill, and it was understandable, therefore, that he decided to hit them with everything he had.

This is a quandary, in fact, in which locust control workers often find themselves. I have a vivid recollection of another occasion when a combined air and ground operation left hoppers dying by the million. Before it began, a large flock of white storks had been feeding heartily and it is just possible that left to themselves they would have cleaned up this particular infestation, which was not

---

* He has, I am sorry to say, since died, partly, I believe, as a consequence of repeated exposure to the extremes of the Ethiopian climate.

very big. But this was a risk which could not possibly be taken. Within three or four days the colony would have been ready to fly as a swarm, able to produce more swarms and so spread the plague. As we made ready to go to work on them the storks moved off to watch from a distance. When we departed they no doubt moved back to a feast which by then would have been drenched with poison. I cannot say that they became casualties, because I was not able to stay to see; but obviously they were at risk.

The use of dieldrin has been banned or restricted in many countries because of its persistency and toxicity to birds and mammals. It belongs to the cyclodiene group of chlorinated hydrocarbons in which other familiar names are aldrin, heptachlor and endrin. Naturalists were the first to become concerned about their effects on wildlife and other mammals and their fears have spread to ecologists, many of whom believe that they may lead to environmental contamination. Why, then, should dangerous poisons of this kind, suspect as they are in developed countries, be used in underdeveloped ones; and in particular on locusts? The answer to the general question, given by some apologists in connection with similar fears which have been expressed about DDT as an agricultural pesticide, is that in order to feed the world population it is necessary to use it as a proven means of protecting the crops, even if this means taking a calculated risk.

This is a reasonable explanation as far as it goes, for neither dieldrin nor DDT have as yet been shown to cause direct harm to man and it is obviously better to take a one-in-a-million chance of their doing so than to have millions starving for want of their protection. As for the use of dieldrin in locust control, it is the very quality of persistency that makes it so useful a tool. It not only kills locusts, on the whole very efficiently, through contact action and stomach poisoning, but goes on doing so for a long time. Sprayed on foliage it normally remains effective for several weeks – a fact particularly valuable in barrier spraying, against hoppers, when swathes of poison are laid down at intervals in order to intercept the insects as they march. The use of techniques of ultra-low-volume spraying, as I have previously pointed out, also greatly reduces the quantities needed to obtain a good kill.

Nevertheless, some doubts are bound to persist. Dieldrin is a general pesticide which destroys not only the insect it is desired to control but other beneficial ones whose existence may not even be suspected. Although the Desert Locust has the great gift of being able, very often, to forestall its insect enemies by gregarizing, copulating and producing a new generation before they have a chance to appear, it is bound, nevertheless, to meet up with a few in the course of its long travels. The fly, *Stomorhina lunata*, for example, may even migrate with it and, poising itself on a nearby bush, wait for the female locust to lay, whereupon it plants its own eggs on top of hers, enabling the fly larvae, emerging a few hours later, to find a meal ready prepared. Eating their way down the egg-pod, they usually destroy it totally; moreover, because the new adult flies emerge before other locust pods are hatched, they are able to repeat the same process, with the result that, in some cases, a whole egg-field may be destroyed. Unfortunately, *Stomorhina lunata* is not found everywhere and notably not in West Africa. Even though only patchily present, it is one of some half a dozen insect enemies of the locust presenting a form of natural control of some importance. Yet a non-selective pesticide is bound to kill the predators with the prey.

Although this by itself is not a sufficient reason for restricting the use of dieldrin, one is bound to speculate about its long-term effects on natural life in general.

The reasoning that, because it is used comparatively sparingly in widely separated places spread over huge areas of country, there is nothing to fear, particularly needs re-examining. As in the case of Alem Seghid's misadventure with the goats in the Danakil desert, there is always the danger when working under pressure that even a skilled technician is liable to be tempted to 'overkill' an infestation of hoppers or swarming locusts by swamping them with everything available. The Anti-Locust Research Centre issues strict warnings on this point, that no area should be sprayed more than once (meaning in the course of the same operation) and that formulations should be very carefully chosen to suit the method of spraying. It is over-optimistic to believe that such instructions, or even those on labels or leaflets coming with the pesticide, are likely to be strictly observed by the technicians in developing countries whose English is imperfect or non-existent. Much of the outcry against dieldrin and other persistent pesticides in highly developed countries arose largely because farmers who were capable of reading and understanding printed warnings either ignored or underrated them. One can hardly expect people in primitive peasant communities to do better. Yet the more sophisticated the methods of spraying used, the more important it is to use them correctly, because of the great areas that can be covered. In a recent West African air-spraying operation using a strong (twenty per cent) formulation of liquid dieldrin the quantity distributed over a total area of 500 square kilometres was six and a half times the recommended maximum when calculated by the acre, and there is no reason to doubt that this happens regularly. The possibilities of environmental damage can be judged by the fact that under controlled conditions in English agriculture a single application of dieldrin against carrot-fly has been found to be so persistent that it is able to give a high degree of immunity for up to ten years; it also, reports Dr Kenneth Mellanby, Director of the Monks Wood Research Station of Britain's Nature Conservancy, 'kills many other soil insects; beetles which are predators on other insects are among the victims'. This may seem of little consequence in desert and steppe where there is little cultivation of economic consequence, but the fact is that we do not know very much about the soil ecology of this kind of country or what the consequences are of upsetting it. We do, however, know that these are very unstable environments indeed, liable to wind and water erosion on a scale inconceivable to anyone familiar only with the green and gentle scenes of countries like England; and we know also, or are beginning to do so, that erosion in any form is not only a result of climatic extremes but is closely linked with the nature of land use and abuse, including use by animals.

The discovery of plentiful and varied animal remains in the Sahara has proved that there was at one time a considerable amount of wildlife supported by much more perennial vegetation than now. It is usually supposed that climatic changes brought about vegetational changes which in turn produced desert; but it is also certain that a very large extent of the Sahara has been man-made. Thane Riney, a distinguished ecologist on the staff of FAO who has closely studied the marginal lands of Africa, is of the opinion that the northern deserts, including the Sahara, are advancing southward at the rate of a mile or more a year along a front of several thousand miles as a direct consequence of grazing by vast herds of camels and goats which have largely ousted the former wild graziers, the now-rare gazelles and oryxes killed by ruthless hunting.

Now, as the Governor of Dankalia told Alem Seghid, it does not matter at all if a few individuals in these excessive herds fall victim in the anti-locust campaign, but it does matter very much, and not only for sentimental reasons, if the remainder of the gazelles and oryxes vanish, for they are part of a crumbling desert eco-system which ought, if possible, to be protected and restored. At a time when there is much talk of reclaiming the desert these are the creatures that can help us do it, if only by acting as environmental indicators. To put it more simply, while they remain we know that there is at least some hope of extending life in this hostile land. When they are gone, more goats and camels will for a time replace them, eating up the remainder of the perennial grasses and shrubs, and when this protection is lost the soil will become total desert. Wind and periodic rain will erode it to a point where it will be irreclaimable. Two-thirds of the Sahara, eroded down to bare rock or gravel plain, is at this stage already. Brave attempts at tree-planting, using new techniques of soil fixing, have shown here and there what could be done given the will; but the plain fact is that while one such scheme is being set afoot ten thousand times as much marginal land is being lost. Only rare patches of the Sahara, outside the oases, support a frail complex of natural life – a few species of trees and shrubs, a few perennial grasses and herbs and a sparse population of birds, lizards, insects, rodents and larger animals which must be able to move elsewhere when the conditions demand it. Few as they are, all these elements, gathered together in a single wadi which is perhaps a hundred miles from another, interact in direct or indirect support of each other. Among these are the locusts; any too drastic methods of wiping them out are liable, therefore, to damage the whole complex. Yet it is in these isolated patches, of course, that much of the spraying, particularly against hoppers, must be done. After heavy rains they may constitute ten per cent of the surface in a given region of desert and it is these, not the desert as a whole, which are in consequence dosed with dieldrin or whatever chemical is used. With the narrowing down of the locust's most-favoured breeding areas we are in fact narrowing down the areas where repeated spraying must be done if the locust is to be kept under control. It is this which is perturbing.

So much hysteria – not confined to the 'antis' – has been generated by the question of poison sprays that it is difficult to keep a clear head about it. Intensive and world-wide lobbying on the one hand by the chemical industry, understandably worried about the future of its very expensive investments, and some conservationists' furious demands for unqualified bans on the other, have hardly helped. The economic quandaries of most of the developing nations are certainly often overlooked. All of them need the best protection they can get in order to feed their enlarging populations; but, because of their adverse trade balances, they must be able to buy it in the cheapest available form. The cheapest and probably the least harmful of the pesticides in general use against locusts, gamma-BHC (Lindane), accounts for about two-thirds of the total quantity sprayed in the average heavy campaign; but in the formulations commonly used it is less than half of one per cent as toxic as dieldrin and so requires the flying of more sorties to achieve the same result. In difficult country, where bringing up supplies of fuel and pesticides presents major logistical problems, dieldrin, though about three times as costly, is therefore often used instead. Over the Desert Locust's natural region as a whole it accounts for some thirty per cent of the use. In the autumn campaign in West Africa in 1968 it was virtually the only liquid pesticide used. (Yet it was this region which, two years later, saw another flare-up.)

In practice, therefore, the choice of anti-locust pesticides possible in a poor country has lain between these two, and it is likely that they will be used in much the same proportions for quite a long while to come. To object to the spraying of dieldrin under these circumstances would therefore be impractical and inhumane. Yet an alternative should be found if possible. One new chemical of promise is DDVP (dichlorvos) – known in one formulation as Vapona. As manufactured now, its persistence is practically negligible, its toxicity to mammals low and its killing power on locusts (also, of course, on other insects) extremely high. Used on swarms in Morocco it knocked them down at such an extraordinary rate that experts considered it to be at least the equal of dieldrin and in the long run no more costly. It is the initial cost – over three times as much as dieldrin – which is the main drawback for nations which might have to stock it for years before needing it. Yet if the governments of developed countries really mean what they say about the preservation of environment, they could surely help by offering such a comparatively safe chemical as part of their aid programmes. (During the 1968 plague the Desert Locust Control Organization of Eastern Africa received a donation of fifty tons of aldrin powder. It is still, as far as I know, unused, which is just as well, seeing that aldrin is transformed in the soil into dieldrin and, if used in the quantities normally employed in dusting, would be deadly. Gifts of this kind really amount to dumping.)

Beyond the search for a safe general insecticide lies the hope that sooner or later we shall have one which is specific to the locust; but whether it would pay the chemical industry, as at present organized, to produce it is more than doubtful. The investment required for the production of a new insecticide has been put at between six million and eight million dollars; to get this back and make a reasonable profit requires continuous unchallenged sales for several years. Obviously the wider the range of pests killed by a general insecticide the greater will be its sales, and the quicker will the investment be recovered. It is for this reason that the manufacturers have reacted so strongly to the threatened bans or restrictions on the use of DDT and dieldrin. The production of specific insecticides, if they can be found, would be just as costly and the market, of its nature, very much smaller. The manufacturers are hardly likely, therefore, to tumble over each other in their efforts to achieve one. The needed research, evidently, will only be done on a sufficient scale if it is made worth the industry's while, that is, by national subsidies in the producing countries or by forming a consortium backed by an international aid organization. The only aid-giving countries with experience of the locust problem in their ex-colonies are Great Britain and France. Britain is already spending a matter of £100,000 a year on locust research and is not likely to give more. France, which used to give substantial financial aid to the West African anti-locust organization, OCLALAV, has recently cut it down. The development of a pesticide designed specifically for the locust could in any case only be part of a wide-ranging search for a variety of chemicals for similar use on a variety of pests; and this would be very expensive indeed. Yet the benefits would be immense, not only to developing nations but in developed ones, where pest control is now a formidable problem. The obvious, indeed probably the only, body with the means to initiate and underwrite research for this purpose is the United Nations. In its application to locusts this would be a logical development of the Desert Locust project.

How then can the lessons of these last campaigning years be summed up?

One returns to the fact with which this baleful and enigmatic insect so often confronts us, that in spite of the intensity of the study concentrated on it (more than on any other insect) we are still, in some important respects, in the realm of the 'Don't-knows'. We are not sure of the long-term consequences of the pesticides we are using against it, nor to what precise extent the 1967-8 outbreak was brought to an end by human agency or what other factors may have helped. We are not even sure, for that matter, whether it did positively end or whether undetected gregarious populations did not survive somewhere in the Sahara to produce, in Mali, toward the end of 1970, a situation uncomfortably similar to that of 1967. It is only possible to repeat, in fact, that as a result of the experiences of the past three years we can be reasonably sure that the Desert Locust can be controlled and its ravages limited, provided there is full international cooperation. This can take a number of forms, including the always valuable aid of non-government organizations such as the Swedish SIDA, whose sponsorship has at last made it possible for FAO to carry out field investigations of pesticide residues in locust habitats. These tests are to be carried out in eastern Africa and will perhaps answer some of the questions I have raised in this chapter.

The multi-million-dollar United Nations Development Programme Desert Locust project launched in 1960, ended in 1970. Thanks to it, there are new field research stations strategically located throughout the regions which have been so often invaded. The guide-lines for this research will continue to be laid down by FAO in conjunction, it is to be hoped, with the Anti-Locust Research Centre in London, where anti-locust workers of many nationalities have been trained under UNDP-sponsored fellowships and scholarships.

The financial ways and means provided by the UNDP for the improvement of the Desert Locust Information Service have enabled all the forty-two countries in the project to know from month to month what the locust situation was like elsewhere, how it was expected to develop, whether and when it would be likely to affect them and what measures they should take. (Because this information is provided by a third party, the service has been extraordinarily valuable in enabling nations, which may be barely on speaking terms with each other, to share vital knowledge.) Intelligence about what the Desert Locust is up to depends, nevertheless, on efficient surveying not only in individual countries but in regions spreading into several. Arranging the UN-financed special surveys of known high-incidence breeding areas has called for a good deal of delicate diplomacy on FAO's part. Thus, Iran and Pakistan have been teamed up to scout southern Iran while Afghanistan and India do the same in the complementary breeding area of south-western Afghanistan. Pakistan and India separately look after their own most dangerous outbreak areas. In the southern half of the Red Sea, where so many plagues have originated, there are now four special seasonal surveys involving teams from Sudan, Saudi Arabia, Yemen and the DLCO of Eastern Africa. In the southern Sahara it is no longer necessary, as it was in 1966, for an Algerian team to come to a frustrated halt on the invisible borders of Niger and Mali, for all three countries, together with OCLALAV, have found it possible to work together under the UN's benign wing.

Most of the countries endangered by the locust now have their own internal radio links, using apparatus provided by the UNDP. In an emergency, therefore, control work can be quickly switched around from one district to another and reports from survey teams can now readily be relayed back to the national headquarters, and so passed on to the Information Service in London. But only in one case – fortunately a vital one – is there a direct link between independent countries. This enables the DLCO at Asmara to speak across the sea to the locust control organization at Aden, thus keeping a check on the movements of swarms between southern Arabia and eastern Africa.

Similar international links elsewhere would be a vast help in combating the spread of a new plague. They are, in fact, an essential part of FAO's plan for a regional strategy based on the Desert Locust's pattern of breeding and migration. The first of these groups embraces India, Pakistan, Iran and Afghanistan. A similar body for the Near East has brought together all twelve countries of the region, and although not all are yet members they seem to have achieved a good deal of working agreement. In North-West Africa a sub-committee of the four Mediterranean countries, Morocco, Algeria, Tunisia and Libya, has agreed to set up a commission which will tie their anti-locust activities closely together; in the meantime they have stretched the hand of co-operation to OCLALAV. Like eastern Africa's DLCO, OCLALAV maintains its own autonomy while working closely with FAO. All these regional groups will set up their own trust funds as an additional inducement to internal mutual aid. And the activities of all of them will be coordinated and guided by FAO.

It sounds ideal. Whether it will be so in practice depends to some extent on the Desert Locust itself. If it goes into a long recession there will be a danger that even the most meticulously prepared defensive organization will become rusty, and then the nations will again be faced with a call to panic stations. FAO has itself prepared for this possibility by setting aside a sum of $500,000 which can be called on in emergency. The flare-up in the south-western Sahara in 1970 was damped down with its aid.

But in fact a long natural recession is unlikely, for although the Desert Locust has been shown to be controllable, it has not, and probably never will be, conquered. Its friends the rains will periodically foster it, encouraging the solitary creatures toward a monstrous gregariousness; and its allies the winds of the Inter-Tropical Convergence Zone, will aid it. The Desert Locust is a litmus paper test of man's ability to cooperate. For the forty-two nations of its region it will also always be the enemy at the gate.

Index